Bored and Busy

American University Studies

Series XI
Anthropology and Sociology
Vol. 52

PETER LANG
New York · San Francisco · Bern
Frankfurt am Main · Paris · London

Phyllis L. Baker

Bored and Busy

An Analysis of Formal and Informal
Organization in the Automated Office

HD
8039
.M4
U532
1991
West

PETER LANG
New York · San Francisco · Bern
Frankfurt am Main · Paris · London

Library of Congress Cataloging-in-Publication Data

Baker, Phyllis L.
  Bored and busy : an analysis of formal and informal organization in the automated office / Phyllis L. Baker.
    p. cm. — (American university studies. Series XI, Anthropology and sociology ; vol. 52)
    Includes bibliographical references.
    1. Clerks—United States. 2. Office practice—Automation. 3. Industrial sociology—United States. I. Title. II. Series.
HD8039.M4U532   1991   302.3'5—dc20   90-25698
ISBN 0-8204-1362-3   CIP
ISSN 0740-0489

CIP-Titelaufnahme der Deutschen Bibliothek

Baker, Phyllis L.:
Bored and busy : an analysis of formal and informal organization in the automated office / Phyllis L. Baker.. — New York; Berlin; Bern; Frankfurt am Main; Paris: Lang, 1991
  (American university studies : Ser. 11, Anthropology and sociology ; Vol. 52)
  ISBN 0-8204-1362-3
NE: American universities studies / 11

© Peter Lang Publishing, Inc., New York 1991

All rights reserved.
Reprint or reproduction, even partially, in all forms such as microfilm, xerography, microfiche, microcard, offset strictly prohibited.

Printed in the United States of America.

DEDICATION

To Patricia Louise Baker
and Donald Lewis Baker

To Douglas Raymond Hotek
and Leanne Baker Hotek

ACKNOWLEDGEMENTS

This book was a collaborative effort as well as a very personal one. Without the many stimulating conversations, paper presentations, article submissions, extensive editing, and graduate seminars this book would not and could not have been written. The ideas contained herein are not mine alone but result from interaction within an extensive network of people, their ideals, and ideas. Bennetta Jules-Rosette's commitment, intellectual integrity, and superb teaching were and continue to be integral to my personal, professional, and intellectual development. Kristin Luker and Rae Lesser Blumberg were also major influences in the continuing formation of my sociological imagination. Training by Hugh Mehan and Beryl Bellman in fieldwork and sociolinguistics was integral to the methodology employed in this project.

I cannot forget my colleagues at U.C.S.D. who not only read, listened to, and commented on the papers and drafts which preceded this book, but also helped me through life's struggles. The support from the Advanced Issues in the Sociology of Knowledge group was appreciated and without compare. Many thanks to Sam Combs, Lisa Phaneuf, Pierce Flynn, Harriet Fleisher, Shiela Hittleman-Sohn, Randy Coplin, Alberto Restrepo, Nick Maroules, and Linda Locklear. There were many other people along the way who gave me tremendous support including Sharon Hays, Charles P. Gallmeier, John Hensley, Stephen Ettinger, Gail Huston, Sharon Cohen, Joann Sandlin, Julia Haase, Martin Hansen, and Mary Ellen Wacker. Thanks to all those I forgot to mention.

Special thanks to the workers at Inland Attorneys, Western Bank, Beach Attorneys, and Sporting Goods International, without whom this book would not have been possible.

CONTENTS

1. **Introduction.** 1
   Clerical Workers, Clerical Work, and the Automated Office, 3
   Assumptions of the Book, 5
2. **Social Contexts of the Automated Office** 11
   Macro-Social Picture of Complex Organizations, 12
   Informal Organization, 24
   Conclusion, 32
3. **Historical, Ethnographic and Comparative Accounts of Organizational Designs** 37
   Historical Account of Organizational Designs, 38
   Ethnographic Account of Organizational Designs, 42
   Comparative Account of Organizational Designs, 65
   Conclusion, 69
4. **Office Practices** 73
   Official Office Practices, 74
   Unofficial Office Practices, 79
   Alienation, Control, and Clerical Work, 82
   Conclusion, 88
5. **Bored and Busy at Work: A Semiotic Model** 91
   Semiotic Square, 92
   Clerical Work Style, 103
   Conclusion, 105
6. **Little Victories and Big Defeats** 109
   Findings and Contributions, 111
   Larger Social and Sociological Issues, 114
   Future Research, 115
   A Final Note, 115
   Appendix A: Methodological Notes 119
   Appendix B: Interview Guide 125
   References 129

1

INTRODUCTION

The stereotypical clerical worker is harried, overworked, and underpaid. This notion is reinforced by both common sense suppositions and the findings of most sociological studies. The subordinate position of clerical work has been further exacerbated by rationalization and routinization of the labor process. New technologies have not only affected the nature of changes in the workplace, but also the rate of those changes. This is especially visible in the automated office where machines now afford the capability to directly monitor behavior and provide immediate and antipathetic feedback (Garson 1988). This, in turn, results in clerical workers' increased susceptibility to subordination and disenfranchisement. A picture of machine-bound people exhibiting robot-like behavior is conjured up in our heads. Yet, do these common sense suppositions and sociological analyses really do justice to the clerical worker? Is there more to the drudgery of clerical work and the automated office than meets the eye? The stereotypes of clerical work and the objective conditions of that work are both products and processes in a seldom considered dialectic: that of the mutually contradicting, yet reciprocally reinforcing, formal and informal organizations which surround clerical workers. The intent of this book is to augment common sense and to elucidate previous sociological notions by examining the everyday world of these clerical workers, in other words, taking into account their informal organization and its relationship to its formal counterparts.

Clerical work exists within a historical, organizational, and social context which precipitates an organizational design that constrains and alienates the workers. These constraints are typified by a routinized work process, inadequate compensation, and the lack of advancement and decision-making opportunities. Additional constraints include not being able to define which work to do, how to do it, or when it will be done. Furthermore, the work is often repetitious, characterized by what we might call "speed up" or "machine pacing," not unlike that of a traditional assembly line.

Many social scientists who have studied women and clerical work have focused on the direct effects of new technologies. Sociologists and other social scientists continue to debate the effect of new technologies on occupational patterns (e.g., Hartmann, Kraut, and Kraut 1986; Hunt and Hunt 1986), clerical workers' class position (Crompton and Jones 1984; Nakano-Glenn and Feldberg 1977; Yiu-Cheong 1980), women's subordinate status (Barker and Downing 1985; Davies 1982, Ferguson 1984), and the history of women in the paid labor force (Groneman and Norton 1987; Matthaei 1982; Rotella 1981). Previous analyses conclude that organizational constraints result in degradation and deskilling of work, and alienation from the products, processes, and meaning of work itself. Such findings and analyses are useful in describing wide-ranging social processes and phenomena. However, do the actual perceptions and activities of clerical workers themselves validate those findings? What is the underlying process that creates, maintains, and then re-creates these organizational constraints and their consequences? Why don't the inherent contradictions of this apparently antagonistic relationship precipitate its downfall? In effect, the sociological question then becomes, what historical, organizational, and social mechanisms operate to maintain this sociological phenomenon?

The world of the clerical worker is characterized by external antecedents and constraints. However, we know that people are not mindless robots preprogrammed to respond only to select stimuli from the external environment. When workers participate in the automated office they direct their actions in accordance with organizational controls, but the process is not a passive one. Blumer argues:

Usually, most of the situations encountered by people in a given society are defined or "structured" by them in the same way. Through previous interaction they develop and acquire common understandings or definitions of how to act in this or that situation. These common definitions enable people to act alike. The common repetitive behavior of people in such situations should not mislead the student into believing that no process of interpretation is in play; on the contrary, even though fixed, the actions of the participating people are constructed by them through a process of interpretation (1962:187-188).

Clerical workers are active participants in the creation of their work world, even when that work is stressful, tedious, and repetitive. This raises the question as to how clerical workers define their everyday world of work and activities within that world. This knowledge is requisite to explaining how clerical workers affect and are simultaneously affected by their work world; the empirical findings herein are concerned with how clerical workers construct their individual realities of work and self while at the same time being part of a larger historical, organizational, and social context characterized by forces of externality and constraint.

Clerical Workers, Clerical Work, and the Automated Office

Since the turn of this century the clerical work force in the United States has grown dramatically and switched from a male dominated to a female dominated occupation. Clerical work has become a mainstay of women's employment. As such, any study of clerical workers is, essentially, a study of women workers. In 1900 there were 877,000 clerical workers about (3% of the work force); 24.1% of them were women.[1] In 1950 that number jumped to a little over 7 million (about 12% of the work force); 62% were women.[2] In 1970 there were 13 1/2 million (about 17.8% of the work force) and 73.6% of them were women.[3] In June 1990 there were approximately 18.6 million clerical workers, (15.6% of the labor force) and 79.6% are women.[4] If we calculate just secretaries, stenographers and typists, women comprise 98.1% of that job category and only 54.6% of the supervisors.[5] From this we can see the huge increase in the amount of clerical workers, and the feminization associated with that increase throughout this century.

The activities of clerical workers, as we know, involve the use of words and numbers. The term clerical once referred specifically to members of the clergy. Its use was expanded to include the modern day clerk, or clerical worker, most likely because the clergy of earlier times were the only literate group in society. It is also interesting to note that the medieval clergy not only possessed the special ability to wield power of the written word, but might also be considered the first employees of a bureaucratic organization (Collins 1986:49). Today, clerical workers typically deal with the written word and work within bureaucratic organizations. The July 1990 United States Department of Labor, Bureau of Labor Statistics classified clerical workers as "administrative support, including clerical." This classification includes supervisors, computer equipment operators, secretaries, stenographers, typists, financial records processing, mail and message distributing, and other administrative support.

Increasing numbers of clerical workers now work in automated offices. Nine-to-Five, the National Association of Working Women, estimated that by 1990 forty to fifty percent of all U.S. workers would have daily use of electronic terminal equipment.[6] They also predicted that there will be 36 million terminal-based work stations in the office by 1990, an increase of 36 percent from 1987.[7] It was estimated that by 1992 personal computers would outnumber operators![8] For purposes of this book, an automated office is one in which some form of computerization is used to manipulate words and/or numbers. Such a broad definition is necessitated by the broad scope of people dealt with herein. The clerical workers considered in this study are primarily word and data processing operators, but also included are secretaries, receptionists, and accounts receivable and payable clerks. In terms of this book, what is particularly important in defining an automated office is that these clerical workers use computers as part of their work activities.

Assumptions of the Book

The first assumption of this book is that social reality is a complex phenomenon with many levels. In order to understand clerical work and the automated office we have to be aware of at least three levels of social reality. Those levels are the
1) macro social context,
2) organizational design, and
3) informal organization.

The macro-social context of the automated office is largely shaped by a market society, rationalization, and patriarchy. These influences have rationalized and routinized the labor process. The macro-social context sets parameters for organizational design. In the automated office the organizational design provides the structural basis for the authority of the bosses and the subordination of the workers. Authority derives from the power to manipulate these structures. Organizational design which delineates the work process and authority structures, is the context within which the informal organization operates. The informal organization is where the macro-social context and organizational design are created and re-created. These domains are recursive and mutually constitutive.

The informal networks are contingent upon clerical workers' perceptions and activities. These networks are where the macro-social context, organizational design, and the informal organization of automated offices come together in some very interesting and thought-provoking ways. Clerical workers' practices consist of subordination and resistance. Basically, bosses try to gain control and workers try to get autonomy. The relationship between the different domains of social reality can be analyzed through "office practices" which constitute everyday life in the automated office. The office practices are in a dialectical and reciprocal relationship with the other elements; they create them and are also created by them.

Another assumption of this book is that organizations exist as some kind of activity within the social world of the people involved. We can not talk about organizations, institutions, or social structure as separate from individuals. Workers are not passive—they are active participants in their work world. This means that workers have their own ways of thinking about their work world, along with the larger social contexts each constructs. Organizational constraints must be more than the words which appear in organizational charts or in policies and procedures manuals. They have to be visible and practiced by the workers.

Further, an ethnographic method is the most conducive to locating clerical workers' perceptions and activities. Ethnography is traditionally seen as an important and inherently "interesting and descriptive" anthropological method. It comes under attack for its primarily descriptive qualities and problematic data collection strategies (Sperber 1982, Clifford 1988). But ethnography is more than a method of description, it can also be used for theory building (Clifford and Marcus 1986).

> What the appropriate facts of social theory are and how to represent them combining both interpretation and explanation is thus a current topic of widespread interest that can be posed rhetorically and repetitively in theoretical discourse, but can only be pursued in the doing of fieldwork and the writing of ethnography. This is why ethnography—hitherto widely viewed outside anthropology as marginal in both its practice (mere description) and its subject matter (the primitive, the exotic, alien other) has been appropriated by a number of fields that sometimes recognize anthropology's labor in this vineyard, and sometimes do not (Marcus 1986:167).

In recent years there has been a theoretical and methodological interest in the relationship between macro and micro structures and processes (cf. Giddens 1984, Knorr-Cetina and Cicourel 1981).[9] Through these studies it is hoped the social sciences will come to a better understanding of social structures, interaction, and everyday life practices—not as distinctively separate phenomena but as interrelated and mutually constitutive. Ethnography is one way to begin to explore that relationship (Tyler 1985, Jules-Rosette 1978b).

Introduction

Organizations are a particularly interesting way to look at the relationship between historical, organizational, and social context because organizations embody structurally defined hierarchical power relationships and are closely connected to the political economy. Some studies of organizations have looked at these relationships. In one study of working class British teenage boys, Willis (1977) argues that power relations are reproduced and internalized through the educational system as the "lads counter school culture." Burawoy (1979) takes a different view than Willis. He argues against the notion that consent to hierarchy resides in cultural institutions, and in favor of a "point of production" analysis. Burawoy says that the basis for consent is the organization of the work process into a piecework game which generates commitment and harmony. In her organizational study, Kanter shows how persons occupying lower status positions possess traits such as a rigidity, rule mindedness, and controlling styles which keep them in subordinate positions (1977:6). Through the use of ethnography, we can come to a better understanding of the relationship between social structure, interaction, and everyday life practices.

The fourth assumption put forth in this book is that technology is inherently neutral. The important determining feature of technology is the context within which it is used. In the United States technology is developed and used within a market society, therefore technology takes on features associated with that setting. The features usually associated with a market society center around inherent human inequality. A market society, rationalization, and patriarchy set the stage for organizational constraints in general, and clerical workers' behavior in particular. Technology, then, takes on the features associated with its social context which is embodied in organizational constraints.

Three domains of social reality within organizations—macro-social context, organizational design, and informal organization—parallel each other with certain themes. Rationalization, lack of worker control, and patriarchal ideology can be seen in all three domains. However, clerical workers' perceptions and activities show us a new dimension. From an analysis of the informal organization we can see how clerical workers' perceptions and activities are both subordinating and empowering. Empowerment on the part of the clerical worker is not evident when an analysis only focuses on social and historical context and/or organizational design. An analysis of clerical workers' everyday life augments our common sense suppositions and adds a new dimension to traditional sociological analyses.

Notes

[1] Historical Statistics of the United States: Colonial Times to 1970. vol. 1, U.S. Department of Commerce, Bureau of Census, p. 139-140. The first year in which the "Employment and Earnings" bulletin of the Department of Labor classified by occupation was 1959. In 1960 the topical heading for clerical work was "clerical and kindred workers." In 1970 and through 1980 that heading had changed to "clerical workers." Between 1983 and 1984 statistical clerks and data entry keyers were added as new categories of workers under the broader category of clerical workers. In 1986 the U.S. Department of Labor, Bureau of Labor Statistics put clerical work in a general heading called "Administrative Support Occupations, including Clerical Work."

[2] ibid.

[3] ibid.

[4] "Earnings and Employment" bulletin, U.S. Department of Labor, Bureau of Labor Statistics, July 1990, p. 33.

[5] ibid.

[6] "Clerical Displacement in the Service Sector in Chicago", a publication of Nine-to-Five, National Association of Working Women, Cleveland, Ohio, June 1988.

[7] ibid.

[8] "PC Novice", Oct. 1990, p. 77.

[9] There are dangers as well as advantages in this effort. The advantages of avoiding unnecessary argument by understanding interrelationships between levels of analysis can lead to the substitution of terms used by those working at one level of analysis to explain a particular aspect of the same phenomenon at another level of analysis. Examples of this are when psychological (Bandura, Ross, and Ross 1961; Homans 1950; Merton 1938), biological (Spencer 1886), or mathematical (Phillips 1974) models are used as explanations of sociological phenomena. Such models are metaphoric devices

for analyzing correlations between social facts, but being fictive, they can distort when generalized to other levels. The recognition of the necessity for understanding how social facts are accomplished in everyday life has generated such different micro-sociologies as symbolic interactionism, cognitive sociology, ethnomethodology, and sociolinguistics. These fields have in common an orientation toward highly empirical and data driven theories. Hence, their primary concern is in accounting for data in the "bottom up" development of theory rather than beginning with ungrounded theory and searching for data in support of it (Glaser and Strauss 1967). For example, Cicourel (1964) criticizes the use of mathematical models for doing sociology because the models have structures that impose themselves on the social facts they seek to describe. Likewise, Jules-Rosette (1978b:549) maintains that the methods used to reformulate and present data affect the explanatory adequacy of the method of inquiry. She warns against assuming that the categorization of social facts into the communicative form of our discipline are "objective" accomplishments. Rather, they follow the same reasoning strategies as common sense forms of thought. These concerns are raised primarily because of the problems inherent in interfacing different theoretical levels. At the same time an explicit and well put together research design can overcome these problems. One of the major genres of sociology which has come out of this concern with integration focuses on the structuring properties of social structure (e.g., Jules-Rosette 1981; Knorr-Cetina 1983; Lynch, Livingston and Garfinkel 1983; Mehan 1978).

2

SOCIAL CONTEXTS OF THE AUTOMATED OFFICE

In this chapter we will peer through a historically situated window at forces affecting workers and the work process in the United States. These forces have set the stage for a constrained work environment characterized by lack of worker control and autonomy. Specialization, rationalization, and sexism are some of the processes and structures which result from the historical and social forces at play in the automated office and most complex organizations in the United States. I begin with an analysis of the historical and social context of the office because the macro-level structures and processes set the parameters for the domains of organizational design and informal organization. This analysis contains a discussion illustrating the historical development of specialization and rationalization which leads to a lack of worker control and autonomy. As specialization and rationalization occurred, office jobs split into two occupational ladders. Feminization happened at the lower status and lower paying jobs which turned into what we now call clerical work. The other occupational ladder remained male and turned into management.

An analysis of the macro-social context necessitates a brief look at market economy and bureaucracy as defining social properties of complex organizations. The automated office is one type of complex organization within which a sexist ideology feminized part of the clerical work process while the management-type of work process remained male. I will first analyze macro-social context of complex organizations and then apply a similar analysis to the automated office in particular. However, even though a historically situated picture of the automated office is important, it is also incomplete. Such a macro-historical analysis overgeneralizes between differing work processes and negates the impact of informal organization on workers. Therefore, a discussion of organizational design and informal organization will be added to our picture of clerical work and clerical workers.

A Macro-Social Picture of Complex Organizations

The market society and capitalist economy in the United States, first and foremost, defines the nature of complex organizations. A market society is where, theoretically: all things, including land and labor, are for sale on the market; the price of items for sale on the market are determined solely by the dynamic of supply and demand; all people are formally free to compete on that market; and a liberal state supports the unfettered operation of that market while simultaneously refraining from interfering in it. In a market society, the activities of large groups of people are coordinated without a conscious human plan, and the fate of people and their environment is potentially determined by market dynamics. Unlike any other economic system, the labor of the many is for sale on the market. Those who control the materials, property, and money necessary for production exploit labor for the end of profit in order to maintain a situation in which large groups of people are treated as commodities. Those in control need not only legitimacy but also the means for controlling the labor they exploit.

One mechanism to control labor within the market society is the specialization of the work process. The specialization of the work process involves a loss of control over that process by the worker. A classic example of this process was provided by Adam Smith when he discussed pin makers who no longer made the whole pin but produced the head, or sharpened the tip, or stretched the wire. When labor is thus divided, no single worker is capable of producing the whole pin. Control of the work process goes from the hands of the worker into the hands of management. The specialization of tasks becomes the division of labor in organizations.

Another mechanism to control labor developed as the market society progressed. That mechanism was a tendency toward a steady decline in the amount of skill required for jobs, or what is referred to as a "deskilling." There was a drop in knowledge for, and responsibility required by, those jobs. Braverman (1974) calls these processes the "degradation" of work under capitalism. The more deskilled positions there are, the lower the wages that have to be paid, and the more control management has. As the market society progressed so did a deskilling of the work process, a further separation of classes, separation of "conception from execution," and further alienation of the worker.[1] However, it is important to point out that in the context of this project, the term deskilled means only that there was a loss of control and decision-making for the workers. The nature of work and the skills required for that work has changed throughout the progression of the market economy, but workers are not necessarily deskilled. The work does not require fewer skills just that the skills are fundamentally different.

The interrelated processes of specialization and deskilling are defining features of a degraded work force. As these processes continue, alienation of the worker from the work process becomes inevitable. This alienation necessitates further control of the worker by management. It thus becomes essential for control over the work process to pass from the hands of the worker into the hands of management. This transition presents itself in history as the progressive alienation of the process of production from the worker and presents itself as the problem of management (Braverman 1974:58).

Although specialization and deskilling can be seen as general tendencies, their growth has slowed down in recent years in some sectors of the economy (Attewell 1987). Strict control is becoming less prevalent in some sectors of our economy as we move further away from an industrial society and toward an information and service economy. Given our rapidly changing technological context, management is forced to move toward policies of less direct control and more skill in certain sectors in order to be competitive. The move toward less control is evident in some "high tech" industries. This is because the less direct control there is over the workers the more creative they can be and, therefore, have the necessary room to create and manipulate new technologies. Nevertheless specialization of the work process and deskilling of the workers has been predominant in the historical and social environment of complex organizations in the United States.

For Max Weber, the market society was only one part of the larger process of rationalization. The historical process of rationalization (which was, in part, set in motion by pious Protestants) makes possible a form of authority and organization which is systematic, efficient, stable, and predictable. Bureaucracy is the epitome of this system. It makes possible the control of human activity in a precise, disciplined, impersonal, efficient, and, above all, predictable manner. For Weber, the realm of rational domination extends beyond the economic to include all areas of life.[2] Those who are controlled by the bureaucracy become mere "cogs" in a human machine; human activity is treated as "raw material" to be manipulated dispassionately toward the end of efficiency. Weber argues that bureaucracy is a social institution which is efficient, rational, and stable.

Although Weber argues that bureaucracy is efficient he also sees it as an instrument for socializing relations of power. For those who gain control of the bureaucratic apparatus, it is a power instrument of the first order because it is objective and rational. The impersonal character of bureaucracy makes the mechanism easily configured to work for anyone who can gain control of it. Once it exists bureaucracy rests upon expert training and functional specialization of work. The people in positions of power maintain control by increasing the superiority of the professionally informed by keeping their knowledge and intentions secret. A bureaucratic administration tends to be an administration of secret sessions and hides its knowledge and action from criticism. It is difficult to dispense with or replace such an apparatus of power.

There are a few points of clarification to be made regarding bureaucracy and market society before proceeding. These two social processes, bureaucracy and market society, are closely related but analytically distinct. First of all, rationalization and bureaucracy predated market society by hundreds of years. As Weber said, bureaucracy is an efficient way to organize large scale and complex structures. One can find bureaucratic structures as early as Ancient Greece, Babylonia, and Egypt. Moreover, some of the largest bureaucracies today are in eastern European socialist countries. The Soviet Union's bureaucratized and routinized work process has had an enormous impact since Lenin first realized the potential of Taylorism and scientific management. In sum, market society and rationalization are quite separate phenomena which are interrelated in the United States.[3] Taken together these social and historical processes have wrestled control of the work process from the workers.

A Macro-Social Picture of Office Work

So, how is this macro-social context of complex organizations characterized by specialization, deskilling, and rationalization manifested in office work? When we look at office work specifically, we have to consider sexism as an important aspect of the macro-social domain. Broadly speaking, office work has developed through three major phases. The first phase began with competitive capitalism and lasted until the turn of the twentieth century when monopoly capitalism obtained a strong foothold. The second phase of office work began with the widespread use of the typewriter during the late ninetieth and early twentieth centuries and the subsequent feminization of clerical work. The final phase is dated circa 1960 with the start of computerization and office automation.

The transitions between phases in the historical development of the office were ushered in by technological innovations necessitated by the social and economic circumstances surrounding each historical period. Bureaucratic specialization of the office occurred continually throughout these historical periods as a way to increase production and profit. Management's control over the work process became an important, if not the most important, goal of that rationalization. Moreover the sexual composition of the office changed from male to female—a phenomenon that is both tied to, and distinct from, bureaucratic and economic advances. Work in the office began developing into two distinct occupational ladders, women were delegated to the lower status and lower paying positions. The historical development of office work was dominated by the processes of specialization, rationalization, and sexism.

The Early Office (before 1880) The early office in the U.S. consisted of two or three people who performed primarily managerial functions. Before the Civil War relations between employees and office workers were quite personal. Worker control was possible through this close relationship and a stricter division of labor or hierarchy was not necessary. Clerks were seen as apprentices. Davies (1982:22) points out that trust between worker and owner was a condition for control. Job tasks were wide, varied and often required high skill levels. These skills included bookkeeping, various manual crafts, copying, and proofreading, to mention just a few. Matthaei calls these early office workers successful petty producers (1982:218).

During the middle and late 1800s, clerical work was not strictly differentiated from managerial work and was often an apprenticeship for it. Office work was defined as a man's domain that paid higher wages, required more varied skills, and held a more prestigious position than in the later two historical periods. Clerical tasks in these early stages can be likened to a craft. Braverman describes the work as follows:

> Although the tools of the craft consisted only of pen, ink, other desk appurtenances, and writing paper, envelopes, and ledgers, it represented a total occupation, the object of which was to keep current the records of the financial and operating condition of the enterprise, as well as its relations with the external world. Master craftsmen, such as bookkeepers or chief clerks, maintained control over the process in its totality, and apprentices or journeymen craftsmen—ordinary clerks, copying clerks, office boys—learned their crafts in office apprenticeships, and in the ordinary course of events advanced through the levels by promotion (1974:298-299).

Persons in these positions enjoyed tasks which were under their own control and had better than average chances for promotion to managerial positions. During this time there were not many women working in these office managerial positions. As a matter of fact there were not many women working in the paid labor force in general. Before the turn of the century, women did not do the majority of the office work and the work that they did do held low status and prestige compared to men's work.

The Middle Years (1880 to 1960) Economic needs and pressures spurred the office to grow in complexity and size while its character continued to change. The rapid growth and complexity of the economic structure in the United States spurred a concomitant growth in clerical work. Mills (1951:68-69) said that "the organizational reason for the expansion of the white-collar occupations was the rise of big business and big government, and the consequent trend of modern social structure, the steady growth of bureaucracy." He also added that "the proportion of clerks of all sorts increased: from 1 or 2 percent in 1870 to 10 or 11 percent of all gainful workers in 1940" (Mills 1956:69). Technological advances, particularly the typewriter, were taking place to deal with the needs of the increasing sizes of bureaucracy and burgeoning economy. Clerical jobs did not proliferate until the second phase of office automation with the shift from small entrepreneurial firms to large corporate organizations.

The work process became more specialized, divided, and ultimately, routinized. Along with this routinization a new class of clerical workers emerged having few similarities with the previously privileged sector dominated by men. The routinized office work required less skill and allowed for the hiring of cheap labor—namely women. The organization of the work process for clerical workers in the office came to resemble factory work (Braverman 1974; Davies 1982; Nakano-Glenn and Feldberg 1977; Rotella 1981). Clerical work became skill-specific and opened the door for its entrance into a secondary labor market position, characterized by low skill levels, little job training, low pay, no tenure, and little job security. Instead of one person doing the accounting, copying, and managing of the office, there were now file clerks, typists, and receptionists answering to and following orders from managers. The specialization of the work process meant a further drop in the required skill level and in the length of the training period. The new slots for clerical workers were filled by relatively unskilled labor who would accept lower wages.

The sex segregation of the office which started in the first phase accelerated in the second. By 1930 when 25% of all women were classified as economically active, most women workers were white, and 40% of nonagricultural workers were in the clerical field (Rotella 1981). In the office, the gendered separation of work spheres began to develop into two distinct occupational ladders. The continued development of these occupational ladders led to two labor markets; one was managerial and the other was clerical. One clear example of the distinction between what men did in the office and what women did in the office can be seen in the development of the secretarial position. The less skilled, more routinized, and communicative tasks were lumped together into the position of personal secretary while the higher skilled and valued tasks were given to the managers, accountants, and executives. Instead of the secretary being an apprentice who could move into the managerial positions, the apprentice's work turned into a secretary's work with few promotional possibilities.

Employers were willing to hire women for the clerical slots, despite high rate of turnover, because job training costs were low and women were already paid considerably lower wages than men (Matthaei 1982; Rotella 1981). Women were willing to take less money because they had fewer occupational alternatives. The availability of an abundant supply of cheap female labor provided incentives to adopt mechanized and routine production techniques that used workers with firm-specific skills. The women entering the labor force were middle class, college educated, and willing to work for lower wages than men of comparable education (Matthaei 1982:222). As women were drawn into clerical jobs, the tasks continued to change and clerical work was transformed into a subordinate position within the occupational structure.

Because of women's secondary position in society, they were offered only the lower prestige and lower paying positions in the secondary labor market. The specialization of the work process and the feminization of the occupation were accompanied by ideological notions that men should get paid a "living wage" while women only needed to earn "pocket money." Additionally, women were seen as better suited for routinized labor processes such as typing. Profit dictated the division of the office into two distinct occupational ladders while patriarchy dictated which gender would fill the lower skilled and lower status positions.

The Final Phase (1960 - Today) As the market society continued to expand with bureaucratization flourishing, technological advances were made to keep up with the economic need. The final phase of clerical work started around 1960 with the introduction of computerized office automation equipment. The social, economic, and technical changes from the first phase through the third follow the same pattern. The occupational line between clerical workers and management was more clearly drawn. The specialization of the work process and deskilling of one group of worker continued. This process was evidenced by word and data processing personnel who became subject to keystroke monitoring. The changing and expanding economy dictated a further concentration of knowledge into the male managerial sector of office work, while the female clerical sector tasks became more specialized. The concentration of knowledge and control into a small hierarchy became, as in factory work, the key to control over the work process and the workers.

The first form of computerization within the female dominated sector of office work was the key punch. In this sector, persons in the higher paid occupations developed the computer programs and codes necessary for analysis of the data, while the key punch operator entered data. Use of the key punch machine is characteristically boring and dull with a high level of pressure. This pressure comes from the machine pacing of the work in which the operators must be speedy and accurate while in competition with one another. If an operator is too slow he or she could be fired. There is no advancement for a key punch operator. Once a key punch operator—always a key punch operator.

After the key punch machine came the word processors. These computers can take on various shapes, sizes, and forms. Computer operators run the machines—entering numbers or words. One can often find them in word processing centers where there is a strong centralized authority unit in a single site. Here the work is often factory-like with quotas and with great physical and social distance between the operators and the bosses. Computer operators are particularly vulnerable to the increasing use of the machine pacing of work which is used by office management as a tool for control. This allows management to have an accurate account of the size of work load and the amount done by each operator, section, or division (Braverman 1974:334).

On the other hand, there can be found decentralized word processing where there is more autonomy and, more than likely, a higher skill level. An example of this type of office is that of the personal secretary who has computer equipment. Personal secretaries, on the whole, enjoy higher prestige and higher pay than those in the typing pool. But these positions are not without problems. Kanter, in *Men and Women of the Corporation*, sees the personal secretary's position in clerical work today as characterized by a husband-wife type of relationship. She argues that secretaries are rewarded by various techniques that keep their mobility low. For example, secretaries derive their status from the position of their boss and not their own. Additionally, the secretary is constrained by an absence of limits on the manager's treatment of her. Most of her rewards are a result of loyalty to the boss. Each individual manager uses his or her own discretion in determining how that loyalty is defined. Finally, as the personal secretary becomes more indispensable in the organization, there is a good chance that she becomes less marketable because her skills are specific to her particular firm. However, even though she is stuck, she enjoys high prestige and pay compared to most other clerical workers. The process of specialization and rationalization takes on different characteristics depending on whether the technology has been centralization or decentralized. The use of computers has accelerated the processes that started in the first phase—increasing specialization, rationalization, and feminization.

This macro-social perspective of clerical work tells us that the market society and rationalization specialized the clerical work process and took control away from the worker leading to two distinct occupational ladders.

Women filled the specialized, lower paying, and less prestigious jobs. So we have added feminization as an important process contributing to the nature of today's automated office. As the economy continued to expand and to become more complex, it developed new technologies and routinized positions in order to keep up with the economic need. Women filled these positions.

Weaknesses in the Macro-Social Approach

The description above presents us with the macro-social context of complex organization and office work. But it leaves many unanswered questions about the specialization and deskilling of the office and office work How are specialization, rationalization, and sexism implemented within an organization? Are all clerical workers degraded and routinized? Do they feel degraded and deskilled? Do they like their jobs? If their jobs are so bad, why do they keep doing them? The macro-social picture of the automated office can not answer these questions, partly because the macro-social approach

    1) over-generalizes between organizational designs,

    2) negates the impact the organizational design, and

    3) excludes the informal organization in their analysis.

These problems inherent in the macro-social approach leads to an incomplete picture of the automated office.

There is a tendency among scholars who use a macro-social perspective to make broad generalizations about the nature of work processes and workers. In relation to the automated office, the focus is on the effect of market society and rationalization on the work process and on the worker. The general consensus is that it is in the very nature of capitalism to routinize, deskill, alienate, and degrade workers and the work process. Most of the literature advances this notion as applying to all work processes without regard to any particular work process or the experiences of any particular clerical worker. When this line of argument is used to analyze the automated office, it implies that computerization has routinized all clerical work and deskilled all clerical workers.

One of the most critical problems with this level of analysis is the tendency to overgeneralize between different kinds of work processes and organizational designs including authority relations. Namely, there is the tendency to generalize from factory to office work. This neglects the effect of specific organizational designs on the work processes and authority relations within automated offices. The deskilling argument is a specific area that illustrates problems with overgeneralizations inherent in broad macro-social analyses.

The sociological arguments made by observers such as Braverman propose that mechanization leads to a reduction of skill level required to accomplish a task while permitting the hiring of cheap labor.[4] Part of this argument is that the automation of clerical work made it resemble factory work in its organization. Thus, according to these models, clerical work shifted from the primary to the secondary labor market as it became mechanized and feminized. The continuation of that shift in the office, since the introduction of computers, has led scholars to assume that work in the automated office is now essentially the same as factory work (cf. Downing 1980; Nakano-Glenn and Feldberg 1977).

Such analyses as the deskilling argument put forth by Braverman, Downing, and Nakano-Glenn and Feldberg enlighten us as to mechanisms of discriminatory practices but do not take into consideration different kinds of organizational designs. For example, Hazel Downing (1980) studied how word processing occupations are oppressive to women and are an instance of sexual discrimination. She describes how women are manipulated at all levels in the business bureaucracy and asserts that word processing represents a deskilling of a typist's tasks. Likewise, Nakano-Glenn and Feldberg (1977) refer to this deskilling as the "proletarianization of clerical work." They maintain that complex skills and mental activity are displaced by largely manual skills in the office just as they were in the factory. These writers assume that deskilling takes place as a result of automation without taking into account work process designs or specific clerical workers.

However, Shepard (1971) points out differences that exist between the work process designs in the factory and the office. He says that there is much less alienation among workers in automated offices than those in factories.

He argues that the occupational structure in the office is generally upgraded as a result of computerization.

> The evidence indicates that electronic data processing equipment generally upgrades part of the clerical work force, while almost no down grading occurs. A common theme is that computers are best at performing the repetitive tasks done manually. As routine jobs are diminished electronic data processing creates higher-level jobs filled by personnel who variously contribute to the operation of the computer system (1971:49).

Consequently, Shepard concludes that

> Contrary to a popular prediction, mechanization and computerization have not made office employees as alienated from work as semi-skilled factory workers. Computer operators, office machine operators, traditional-type clerical workers, and software personnel are all less alienated than assembly-line workers, the prototypal factory worker (1971:105).

Generalizations made about the effects of automation are problematic, especially when they confuse different types of work process designs. We cannot assume that all work processes are effected in the same way in every situation..

There are also generalizations which confuse the effect of computerization on authority relations. In the case of office automation, word processing involves the acquisition of highly technical skills in addition to those needed to operate an electrical typewriter. For example, word processing involves the acquisition of the same and similar skills used in other forms of computer work, such as data processing. Both can be highly technical and involve a practitioners' knowledge of all the tasks employed by a typist and much more. In the management sector, individuals have a larger array of responsibilities, whereas those in the clerical work force are much more restricted in their work (cf. Jaques 1978). Nevertheless, the acquisition of technical skills to become computer literate is the same wherever one is within that organization. Computer literacy on the part of management has been viewed as a big plus and not as an indication of deskilling as it has been viewed for clerical occupations. Even though clerical workers usually undergo special training and enroll in extra-curricular education programs to become computer literate and proficient, the assumption continues that word processing represents a deskilling of the clerical trades.

However, some recent national and cross cultural analyses (Attewell 1987; Jules-Rosette 1981) show that mechanization and automation do not necessarily mean that a task is deskilled. Thus, it becomes problematic to generalize from macro-social data to all organizational designs whether it be work processes or authority relations. In the automated office deskilling is variable depending on the work process design and authority relations.

Automation is not inherently negative as many macro-social theorists hold, but is shaped by particular people within particular organizations. Technology does not dictate how it is going to be used, organizations do. Clerical work is an outgrowth of a kind of bureaucracy that is contingent upon the use of technology. Even though the deskilling theories describe important aspects of workplace discrimination, they overly generalize between and among organizational designs, particularly the work processes (factory/office) and authority structures (management/clerical) involved. It is the organizational design in an office within the context of the larger macro-social environment that allocate status, privilege, and prestige to positions and to inherently neutral technologies.

Informal Organization

A macro-social analysis of work also excludes informal organization. This prohibits an understanding of the intricacies of the effect of informal organization on the workplace. Left unanswered in a macro-social approach is the question of how and why are specialization and deskilling reproduced. How are these processes experienced by the workers in the automated office? Most of the macro-historical literature assumes that individuals are passive recipients of culture through historical and social forces. However, workers do not just receive macro-social factors but are active creators of them. Specialization, rationalization, and sexism seem to describe the office, but what happens when we look at the informal organization?

The importance of informal organization as a defining characteristic of companies broke from classical thinking about organizations. Chester Barnard in *The Functions of an Executive* defines a formal organization as a system of consciously coordinated activities of two or more persons (1938:73). The important word here is persons. Although the macro-social context and organizational designs are important aspects of organizations, companies are run by people, not by structures. Organizations are collectivities. Through interaction within these collectivities, the ideas, values, interests, and expectations of individuals form into an informal organization. Participants within complex organizations generate formal and informal norms and behavior patterns which create, re-create, and occasionally modify the macro-social context and organizational designs .

The studies of informal organization started with the human relations school coming out of the Harvard Business School in the 1920s. The human relations model was the first to look at the correlation between interpersonal relationships and productivity. This perspective grew out of a critique of Social Darwinism and scientific management and stressed employees' cooperation and group cohesiveness as keys to increased production. The human relations model wanted to change management's focus on productivity from formal requirements and economic rewards to social rewards. Social rewards, they argued, would be a better incentive for workers to align with management's goals and therefore increase productivity. The human relations model was based on the notion that informal group behavior was fundamental to the smooth running of complex organizations. The human relations school sought to manipulate social variables in order to increase motivation and productivity.

The human relations school got started as a result of studies which were conducted at the Hawthorne Works of Western Electric plant outside of Chicago in the mid 1920s. Studies done at the Hawthorne plant left the researchers baffled. They were originally interested in the effects of lighting on productivity. Changes in degree of lighting increased output in the control and experimental groups. The researchers kept running experiments and in both the test and control groups productivity kept rising. The Hawthorne studies continued with the relay room experiments where such variables as rest periods, length of workday, and pay were manipulated and tested for effect on worker productivity. Each variable was manipulated and then changed back to the original state; productivity continued to go up. The Hawthorne studies also included a series of interviews where the researchers found that, under normal conditions, the informal work group placed restrictions on output. Finally, during the bank wiring room study, there was a specific focus on the informal work group. Once this phase of the Hawthorne studies started there were no variables manipulated. Instead, the researchers gained the confidence of the workers and were recognized as members. From the findings of the previous studies they assumed there would be an increase in production because the workers were being recognized and treated as individuals. However, there was not an increase in productivity in this phase. This was due to restrictions of output implemented by the workers. Because the plant was getting closer to closing down, the workers became fearful of unemployment. Additionally, they wanted to protect slower workers, keep standards from rising, and keep management satisfied. Such sanctions as ostracism, ridicule, and name-calling were used to restrict output. Hence, group cohesiveness was related to productivity but the relationship was complex.

After studying the Hawthorne data, Roethlisberger and Dickson (1939) argued that group cohesiveness led to the increase in production during the first two phases. The start of rest periods in the relay room experiments gave the workers a chance to interact and become friends. The network of informal relations proved to be an important variable. Additionally, Mayo (1945) in an analysis of the same data, described the importance of the research project itself for worker satisfaction. He concluded that because there was an interest in their work and workers were able to discuss issues of concern. This led to the increased productivity and was labelled the "Hawthorne effect." The rise in productivity was a result of some kind of social reward. These clinical studies, coming out of the Industrial Research Department at Harvard University, were the first to focus on the importance of the informal organization in productivity.

There was a plethora of studies published after the Hawthorne experiments. It was hoped that a positive relationship existed between group cohesiveness, morale, and productivity. If there was such a relationship, then management could foster it to their benefit. However there is no conclusive evidence for a positive relationship between productivity and group cohesiveness. Like some phases of the Hawthorne studies, Schacter et al. (1951:235-236) found that group cohesiveness could adversely affect productivity, "in the negative induction conditions, low cohesive subjects are less acceptant of induction and more productive than high cohesive subjects." Group relations can be used for or against productivity. This was obvious in the bank wiring room studies which illustrated that social factors, such as group cohesiveness, relate to the restrictions workers place on output and penalties they place on those who do not.

In sum, the human relations school stressed the human element in organizational analysis and structure. However, the relationship between the informal organization and productivity remained nebulous. Nevertheless, these theories were the single most important historical foundation for the informal organizational approach to complex organizations. Even though the human relations model was based on a management perspective, it illuminates the process whereby workers define (albeit feebly) themselves and their behavior in relation to the formal organization (potentially undermining management's goals, but more often furthering them).

As it became clear that the relationship between productivity, group cohesiveness, and cooperation was complex and nebulous, researchers started looking at autonomy and alienation. They put much less stress on productivity and motivation, instead, they looked at worker satisfaction. In general the studies show that worker satisfaction is higher when there is a greater level of autonomy and responsibility and a lower level of alienation (Blauner 1960, Shepard and Herrick 1972). These studies produced more conclusive results than the earlier studies that had focused solely on productivity and motivation.

For example, the condition of autonomy has been found to be central to the lives of workers. As a matter of fact, the lack of autonomy makes blue collar workers the least satisfied on measures of job satisfaction. One way to measure satisfaction is to find out if given the opportunity workers would keep the same jobs. Results from a survey conducted by the United States Department of Health, Education and Welfare published in *Work in America* said that "Dull, repetitive, seemingly meaningless tasks, offering little challenge or autonomy, are causing discontent among workers at all levels (xv)." Discontent was strongest among blue collar workers, only 24% said they would take the same job. Autonomy is a component of a job which is related to worker satisfaction.

Another condition of worker satisfaction is one's role in the organization. Rosabeth Moss Kanter (1977) talks about the importance of roles and positions for workers' attitudes. She found that for white collar workers in lower level and dead-end positions there were depressed aspirations, low commitment, and bad self image. Workers respond to the powerlessness of their position by becoming apathetic to the organization and not taking risks. Kanter argues that the functioning of organizations is weakened as a result of roles and positions which do not contain adequate responsibility and autonomy. In general, when a job allows for more autonomy and responsibility there is higher worker satisfaction. Attitudes and behaviors are shaped by the fact that people have varying degrees of participation, leadership, and influence in organizations. The more the workers are involved in the company, the better it is for both the worker and company. Autonomy improves work life, satisfaction and, probably, productivity.

Burawoy (1979) does not analyze the automated office specifically, but he looks at factory work and argues that consent to subordination resides at the point of production. He undertook an ethnographic study of Allied Corporation while working on the shop floor. Burawoy argues that the workers' perspectives and activities obscure and secure their subordinate position. The manipulation of the work process by workers into a piecework game allows them to gain self respect through "making out" on the shop floor. This generates commitment and harmony for the piecework system and results in the workers' submission to degradation.

Additionally, Burawoy notes that the internal labor market with its job and promotion structure creates competitive social relations among the workers and promotes a specific ideology of "possessive individualism" on the shop floor. The existence of competition for promotions tends to displace conflict against the system in favor of individual competition between the workers. Finally, the grievance procedures and collective bargaining coordinate the interests of management and unions. This obscures subordination through the constitution of workers as individuals and not as members of the same class. These procedures expand choices for workers and add to the basis of consent for their subordination. Burawoy's ethnographic study of the informal organization of workers sheds much needed light on the notion of subordination in the factory.

The most recent theoretical and methodological perspective on informal organization is an analysis of the everyday life behavior and language use by organizational participants. This perspective is unique because it utilizes an in depth analysis of worksite practices and procedures and does not focus on increasing productivity and motivation. Rather, these studies are interested in the production of knowledge and culture within the workplace. Informal organization is comprised of social and practical aspects of the workplace as understood and lived by the people interacting therein, in the context of, but not determined by, the larger formal organizational structures in which they operate. Many of these studies are in the substantive areas of science and technology. These studies analyze practical actions of workers in a hope to understand how things, like science and technology, create and maintain their spheres in society. I discuss those studies here as examples of a particular framework within the sociology of work literature which focuses on worksite practices as sociologically interesting and important.

For example, this type of study is interested in the behavior that actually occurs during a scientific discovery. Such studies do not focus on what the scientists or officials say happened, but in the everyday life behaviors that went into that discovery. To answer this type of question an ethnographic data collection method is necessary. This approach is unique because the main thrust is that whether the behavior is in a lab, an office or at home, people are the creators of reality. How certain aspects of reality are created becomes important. Moreover, reality is not seen as an identifiable state but is comprised of many factors including multiple knowledge systems.

Garfinkel, Lynch, and Livingston in "The Work of a Discovering Science Construed with Materials from the Optically Discovered Pulsar" asked the question "What does the optically discovered pulsar consist of as Cocke and Disney's night's work?" Cocke and Disney are the two scientists who were in the lab the night of the discovery. Garfinkel, Lynch, and Livingston did sociolinguistic analyses of a tape recording of the night that the scientists discovered the optical pulsar at Steward Observatory, January 16, 1969. Garfinkel, Lynch and Livingston argue that the independent Galilean pulsar is a cultural object, not a physical or natural object. This cultural object is a result of the scientists' night work and their embodied practices. Their work, as locally produced and orderly, created this phenomenon. Before the orderly write up and cultural creation, there really was no pulsar. The practices of the scientists' work provide the properties of the independent Galilean pulsar. These practices are part of an institutionally established membership in a scientific community and are not directly related to the pulsar.

Bruno Latour in "Is it Possible to Reconstruct the Research Process?: Sociology of a Brain Peptide" studied what happened after two physiologists discovered that a growth releasing hormone actually blocked the release of certain growth hormones. After such a discovery it is traditional for scientists to find analogs which could make the amino acids work faster, work better, cheaper to make, work in different kinds of ways, etc. Latour was interested in "...what processes led the laboratory workers to choose the hundred modifications they eventually produced" (1980:55) when there were virtually millions of possibilities. From interviews and observations Latour concluded that the 1) research process is heterogeneous; scientific statements are never pure but come from many parts of the social world, 2) research project is contextual; the amino acid under study meant different things in different labs, 3) research process is opportunistic; people studied different analogs for reasons such as money, 4) research process is idiosyncratic; local conditions of the laboratory are defining characteristics of what gets studied, and 5) research process is fiction-building; that the process is not logical or rational but has to be written up in that way. Latour concludes that work in the laboratory "...is nothing more or nothing less than the reset of our daily world and daily stories of fiction and disorder (1980:69)."

In a similar tradition, Combs, Phaneuf, and Flynn (1985) did a cross cultural comparison of the production of scientific knowledge in ethology laboratories in the U.S. and France. Parallel ethnographic investigations were conducted at the Hubbs Marine Research Institute in San Diego Ca. and at the University of Paris. Ethnographic, ethnomethodological, and semiotic methodologies were employed in an attempt to discover the variety of practical actions employed by animal ethologists in the investigation of animal communication. Findings showed that the discipline of ethology did not exist as an exportable body of theories and methods. Instead, local ethology laboratories constructed their own local definitions of scientific cultures and ways of reporting their respective practical findings. Ethology as a discipline was a situated production by different international groups of researchers attempting to produce discourseable consensus between themselves.

The findings of these types of studies put an end to an age old assumption that science is an orderly straightforward process. Garfinkel, Lynch and Livingston (1981), Latour (1980) and Combs, Phaneuf and Flynn (1985) find that the practices which make up science in these laboratories are chaotic, illogical, opportunistic, contextual, and constantly reconstructed. All were interested in how science was not an orderly activity but a social construction. Scientific knowledge is essentially a cultural object. These studies all have in common a focus on the practical action in everyday life behavior and worksite practices. They are interested in the taken-for-granted assumptions and the routines of the workers.

The informal organization is an important element of any complex organization. The impact of social and practical actions and behaviors affect the workplace and determine the nature of that workplace as well as organizational design and the larger macro-social environment. The everyday practical actions of workers create and re-create what actually happens in the workplace, as well as in the larger society.

Macro-social sociology discusses the processes through which the push for profits and rationalization have been the catalyst for the routinization, deskilling, and feminization of clerical work. However, they stop short of telling us about meanings, practices, and experiences of the participants. Studies of informal organization allows for an analysis of how social and historical processes are reproduced and experienced by the workers. It points to a dialectical relationship between the macro-social context and informal organization. The dialectical relationship simultaneously creates and maintains specialization, rationalization, and sexism.

Conclusion

The origin and reproduction of workplace subordination is possible only through a combination of organizational variables that are shaped in complex ways by the market society, patriarchal ideological system, and worksite practices as they have evolved over long periods of time. The specialization and rationalization of the work process and of the workers was conducted by the economic system through bureaucratic rules, regulations, impersonal treatment, and the division of labor. These processes are the basis for organizational designs of today's automated office. Organizational design including authority structures and the work process originate within this macro-social context and serve as instruments to control the worker and make profit. The macro-social context of organizations impact companies, but these companies are really only collectivities of people. These collectivities engage in activities that re-create and maintain organizational constraints. The relationship between the macro-social context, organizational design, and the collectivities is, therefore, reciprocal and constitutive. They are domains of social reality which contain parallel processes. Historical and structural sociology is not wrong but incomplete; it does not go far enough. In order to understand today's automated office it is necessary to look at a variety of factors in addition to broad economic, social, and historical trends. A micro-analysis allows for a deeper understanding.

Notes

1   One of the mechanisms that is used to deskill is the concentration of knowledge into the hands of a few. This constitutes a further division between the classes and creates a separation of "conception from execution," one of Braverman's most important theoretical contributions. He explains it as follows:

> The breakup of craft skills and the reconstruction of production as a collective or social process have destroyed the traditional concept of skill and opened up only one way for mastery over labor processes to develop: in and through scientific, technical, and engineering knowledge. But the extreme concentration of this knowledge in the hands of management and its closely associated staff organizations have closed the avenue to the working population. What is left to workers is a reinterpreted and woefully inadequate concept of skill: a specific dexterity, a limited and repetitious operation, "speed as skill," etc. With the development of the capitalist mode of production, the very concept of skill becomes degraded along with the degradation of labor and the yardstick by which it is measured shrinks to such a point that today the worker is considered to possess a "skill" if his or her job requires a few days' or weeks' training, several months of training is regarded as unusually demanding, and the job that calls for a learning period of six months or a year—such as computer programming - inspires a paroxysm of awe (1974:443-444).

2   Sharon Hays was integral to the analytical discussion of the interaction between capitalism and bureaucracy put forth in this chapter.

3   These analytical distinctions were brought to my attention my Rae Lesser Blumberg during our many discussions about technology, work, and the change from feudalism to capitalism.

4 Attewell, in his paper, "The De-skilling Controversy" offers a series of theoretical, empirical, and methodological criticism of the deskilling position, paying particular attention to contemporary trends in computerization of clerical work. He essentially shows that there are several theoretical arguments which explain why the profit motive does not necessarily drive managers to deskill their work force. He says that deskilling has not been the master trend of occupational change in the twentieth century. For a thorough critique of Braverman see Attewell (1987). He does an in depth criticism of Braverman's work on deskilling. The following six points highlight his criticisms.

1) He points out that automation could cause a disproportionate eradication of low skill level jobs and offset a decline in the skill level of the economy as a whole.

2) Attewell also criticizes Braverman for only focusing on capital logic. Capital logic is the overwhelming concern under capitalism for the generation of profit. However, at a certain point a further division of labor is not longer efficient. Therefore, deskilling would lower profit and would not be in the best interests of management.

3) Braverman seems to generalize from the entrepreneur to the economy. However, there are many intervening variables affecting the economy which do not effect the sectors and change in the level and kinds of imported goods. This creates different demands for labor markets and effects demands for skill levels.

4) Another point that Attewell makes is that newly created low skill level jobs could be promotions for some people. At an individual level deskilling is problematic.

5) Deskilled jobs do not always translate into deskilled people.

6) Attewell argues that Braverman underestimates the amount of skill required to do some jobs. In other words, if we look closely at some of the jobs considered deskilled, they may require higher skill levels than previously thought.

3

HISTORICAL, ETHNOGRAPHIC AND COMPARATIVE ACCOUNTS OF ORGANIZATIONAL DESIGNS

> In anthropology and all other human sciences at the moment, "high" theoretical discourse—the body of ideas that authoritatively unify a field—is in disarray. The most interesting and provocative theoretical works now are precisely those that point to practice (see Ortner 1984), that is, to a bottom-up reformulation of classic questions, which hinges on how the previously taken-for-granted facts of high theory are to be represented....The concepts of structure on which such perspectives depended are really processes that must be understood from the point of view of the actor, a realization that raises problems of interpretation and presents opportunities for innovation in writing accounts of social reality (Marcus 1986:166).

This study is concerned with clerical workers' perspectives and activities—their everyday life. In order to understand their everyday life it is important to analyze the many different levels of social reality which affect it. The macro-social context discussed in the previous chapter provides parameters for the organizational designs of companies. Within the United States, these designs have created a constrained work environment characterized by specialization, rationalization, and sexism. This chapter describes how these macro-social patterns get played out at the level of organizational design in specific companies.

Our journey into the everyday life of the clerical worker begins with a look at organizational designs through their application in specific authority structures and work processes. In order to understand clerical workers' perspectives and activities it is important to understand the contexts within which their perspectives and activities take place. This book is based on an ethnographic study of four automated offices in San Diego, California. First, I will present a brief and general account of the historical development of authority structures and work processes as they relate to control in the office. Second, I will give an ethnographic account of the authority structures and work processes of each firm in this study. Finally, I will present a comparative account of the organizational designs as they relate to workers' satisfaction.

Historical Account of Organizational Designs

Authority structure is an important component of any organization. Over the past century, the structure of authority relations in U. S. companies has changed dramatically. We have witnessed twentieth-century trends that have moved from a position of strict control, characteristic of the factory system, to more flexible and democratic control mechanisms. Although this is certainly not the trend in all parts of the private sector, the switch from manufacturing-based to information-based industries has necessitated changes in control and therefore in the structure of authority relations.

Around the end of the nineteenth and early twentieth centuries with rapid capitalist expansion and a ready supply of labor, Social Darwinism was the main theory of industrial relations. Industrial practices were characterized by strict control and the notion that only the "fittest will survive." If the worker could not take the pace of the factory he or she could be easily replaced. The treatment of the workers was ruthless. The ethic was one of individual merit and the message was clear; only the strong should or could survive in the competition which was considered natural and inevitable.

Historical, Ethnographic, and Comparative Accounts

Success or failure of the worker depended on how fit he or she was in the race for survival.

The response to this rough and strict theory of industrial relations was the rise of the labor movement and unionization. After the structure of apprentice-journeyman-master tumbled due to the rise of competitive capitalism, bosses became less knowledgeable than the workers about the work process. After a time the workers began to doubt their employers' authority and protested because the workers knew more about the work than the bosses. The authority of the bosses was not as encompassing as the authority of the master.

The rise of violence associated with the establishment of unions was a result of hard line management techniques and lack of knowledge by the managers. Unionization brought to light many problems plaguing the factories and offered solutions to those problems. However, management was not willing to fill all the needs of the unions because of their fear of losing profit and control. Compromises were reached in the form of shorter hours, better working conditions, retirement systems, and ranked promotions.

Since the time of the Social Darwinists and the rise of the labor movements, there has been considerable debate about the effectiveness of centralized versus decentralized authority structures. The argument for centralized authority structures is that strict hierarchical control can be used to negate the tendencies of the workers to pursue their own interests and is more economical because it attacks problems in a coordinated fashion. Large organizations can keep better records of accounts and clients through the centralization of information. A centralized hierarchical structure for authority relations can also develop more encompassing solutions to organizational problems and create equitable distribution of rewards.

Arguments for decentralization emphasize adaptiveness and innovativeness and argue that loose coupling between organizational levels is the most efficient way to be maximally responsive to customer demands. Decentralization can break up power alliances in an organization and make for a more effective organization. Moreover, if the workers have more autonomy, there will be a higher level of worker satisfaction and therefore higher productivity and creativity.

Since the 1940s there has been a move toward decentralized authority structures. Worthy, in "Organizational Structure and Employee Morale" concluded that "flatter, less complex structures, with a maximum of administrative decentralization, tend to create a potential for improved attitudes, more effective supervision, and greater individual responsibility and initiative among employees. Moreover, arrangements of this type encourage the development of individual self-expression and creativity which are so necessary to the personal satisfaction of employees and which are an essential ingredient of the democratic way of life" (1950:179). The move toward flatter authority structures has taken place since society has been moving away from an industrial production mode toward an information industry which has higher levels of uncertainty. Organizations are adapting to such changes in the economy. Peters and Waterman, in their best seller *In Pursuit of Excellence*, argue that the most successful businesses in the United States are those which use a "fluid" type of management approach and stress informality among the staff. Although classical approaches are still very evident today, the increase in the need for creative workers who have the ability to handle change brings a move from a traditional strict authority structure to a flatter and more autonomous structure. Where there is a less routinized work process there is less need for strict control and a more humanizing environment.

However, such changes do not indicate that control over the worker has been given up. Organizational constraints still exist. Companies are utilizing bureaucratic rather than autocratic control. Although bureaucratic control implies strict formal rules and lack of creativity, when compared to simple or autocratic control, it gives the worker more autonomy. When there is more slack given to the workers, they may set higher standards than management. This creates a motive for management to loosen their supervision of the innovative and educated worker in order to increase productivity (Perrow 1970). So there is a tension between maximum bureaucracy, maximum control and creativity in a highly changing technological environment. Parts of corporate

America are adapting to the changes by using less hierarchy with more bureaucratic, not autocratic, control for the purpose of increasing profit.

The design of the work process is another important element of organizational design which has an impact on the worker. The work process has gone through many changes that are closely linked to changes in the economy, technology, and authority structure. As competitive capitalism took a foothold and control issues came to the fore, "scientific management" became a dominant force in the United States.

Scientific management was developed early in the twentieth century by Frederick Winslow Taylor. Scientific management was premised on the notion that careful study of the requirements of each person's job guarantees effectiveness of an organization. In the hands of management, these requirements can be planned into tasks with close supervisory control in order to guarantee the most efficient way to produce goods. Job techniques were redesigned to make optimum use of human abilities. Under scientific management, wages were reduced and ranks were split. Each work day was fully planned out in advance with all tasks specified. What scientific management tried to do was remove all discretion on the part of the worker. This process is rationalization at its best.

Scientific management was based on the notion that employees work as slowly as possible. The worker needed special incentives, e.g., higher wages, premiums, bonuses, etc., so when workers gave more, they would get more. Cooperation between labor and the capitalist, according to Taylor, brought success for the capitalist in the form of profit and for the worker in the form of higher wages. If workers became interested in increasing surplus, then the factories would be more successful. However, labor unions in the United States had power and the workers were able to see that productivity, with or without higher wages, was helping the owners more than themselves. Additionally, under scientific management, the blame for production problems went to management because they were supposed to keep up production. But, management did not want the blame for "lazy" workers. As it turned out, scientific management was not efficient, particularly in a post-industrial system, because the workers were not able to solve production problems or deal with the necessary changes that arise under advancing technology.

The historical development of authority structures and work processes has centered on control over the workers. That control shifted from a strict, autocratic control to a more bureaucratic control that is evident in both authority structures and work processes. In the world of new and rapidly developing technologies, organizational designs require flexibility and adaptability to change while sacrificing classical authority and work process structures.

Ethnographic Account of Four Organizational Designs

What follows is an ethnographic account of four automated offices. The methodology employed for this research was a combination of qualitative methods using observational, participant observation, and in depth interview data collection strategies. Observational studies were conducted at two San Diego word processing centers; a bank (Western Bank) and a large law firm (Inland Attorneys).[1] Participant observation was done as a data entry clerk in a law firm (Beach Attorneys). In depth interviews were conducted with persons involved in a multinational corporation (Sporting Goods International). During these stages of the research project, the focus was on relationships between management and workers, social organization, and the cultural world of clerical workers. The ethnographic focus was fortuitous because it opened avenues for analyzing the effects of organizational design on the clerical workers' perspectives and activities. For example, the participant observation phase of my data collection was useful because it allowed me to get at the nuances involved in the setting. Most importantly, I was able to experience the effect of organizational design as a clerical worker in the automated office. This facilitated locating connections between the macro-social context, organizational design, and clerical workers' perspectives and activities.

This account looks at how the historical development of organizational design gets played out at the level of specific companies. Clerical workers' behavior occurs under organizational designs which wrestle control from the workers. This condition remains fairly consistent throughout the four settings. Organizational design, in the form of authority structures and work processes, can be seen as strategies by management to maintain control and reflections of the larger social and historical processes of specialization, rationalization and sexism.

Inland Attorneys

"Inland Attorneys" is one of San Diego's largest law firms employing 42 attorneys and 70 support staff. The firm is located in a major financial district in close proximity to downtown San Diego. They do a wide range of legal specialties including corporate law, water agreements, insurance claims, and bankruptcies. Inland Attorneys is housed in a three story office building with word processing centers on each floor. The floors, as well as the word processing centers, of the firm were divided by law specialties. Inland Attorneys is held in high regard by the law community and is growing rapidly.

I obtained entry into Inland Attorneys through their personnel director, Connie. We were introduced by a consultant who organizes and leads stress workshops for people who work in computerized offices. I developed and submitted a research proposal requesting to observe their word processing centers for a month, at different times of the day. I also requested interviews with the operators during lunch or after hours. I offered to submit a final report of my findings to the company.[2] Connie took my proposal to the Board of Directors for final approval. The Board was concerned about whether I would cause any interference to the workflow. In order to convince them that I would not interfere, I had to cut down the number of hours I would be there and interview only on the operators' time. The proposal was approved and followed except for a few extra hours of observation with the consent of those involved. I interviewed three supervisors and four operators. I conducted 16 hours of observation.

Authority Structure The word processing centers had the responsibility for processing any documents over three pages long. There were personal secretaries who worked directly for the attorneys. The secretaries typed letters, set appointments, answered phones, and did other clerical and personal tasks for their bosses. There was a separate department for the data processing needs of the company. I had access to the word processing centers but not to the secretaries or the data processing department.

The computer operators at Inland Attorneys were directly involved in an authority structure comprised of four levels. The four levels were the

1) attorneys assigned to specific centers,
2) their personnel director,
3) coordinators of the centers, and
4) computer operators.

The division of the word processing needs into three centers made the authority structure relatively decentralized, except for the coordinators on each floor.

At the time of my data collection, there were seven word processing operators at Inland Attorneys all of whom were accountable to Connie, the personnel director. Connie also had responsibility for the supervision of the rest of the support staff including secretaries and data processing personnel. Because she was the administrative boss, Connie made most of the decisions regarding hiring, firing, equipment, purchasing, etc. She did not engage in much direct interaction with the word processing operators. The operators felt alienated from her and resented her managerial style.

My liaison with the firm was Christy, Connie's administrative assistant. Christy was supervising the operators while I was at Inland Attorneys. Over time we developed a good rapport. Christy would meet me in the reception area and escort me to the centers for observation. The operators liked her. She was a very pleasant person and liked her job which carried many responsibilities.

Each of the centers had two or three operators, one of whom was a coordinator. The coordinator was responsible for the operators and for the work going into and out of the center. She was the formal liaison between

Connie, the attorneys, and the operators. Although there were times in which Connie or the attorneys would talk directly to operators—when Connie was criticizing their work or the attorneys wanted something processed in a special way—there was not much interaction between the levels of authority.

All of the operators and coordinators were women. Most of them were highly trained and had previously been legal secretaries who were hired from within the firm at the time of automation. The coordinators were responsible for starting up the machines, backing up the data, distributing the workload, and supervising the operators. The people with the most seniority were the coordinators. The operators were required to know the word processing software and to do the tasks given to them by the coordinators. The tasks were primarily word processing legal documents.

The first floor coordinator, Brenda, was autocratic and played a classic boss role. Ernestine, the operator on that floor, had to ask permission to go on break and was not allowed to make personal phone calls. There was little interaction in the center. Brenda felt that talking between operators interfered with the work flow. She said that in "her" center "we try and work quietly because I like the documents to come out as final drafts the first time." According to Brenda the first floor center was more productive than the other two. During the period of my observation she kept glancing at me and was very concerned that I not see her making mistakes. She was quite nervous and commented that she did not usually make so many mistakes. I had not seen any mistakes. Brenda was the only coordinator who asked if I wanted to see the production reports which were used to keep track of how much and how quickly the work was done on each floor. Ernestine was quiet but rebellious. She was conscientious and a good operator. However, she had problems with Brenda's strict supervisory style. Ernestine played by the rules while working but once she was on break she would complain about Brenda.

Andrea was the coordinator on the second floor. She was much less autocratic than Brenda. She had worked in large centralized centers before and liked working at Inland Attorneys because they had a decentralized structure. Andrea said the decentralized structure was better because "the girls are familiar with the attorneys, secretaries, and the type of documents so they can be fast and more efficient." Andrea was much more relaxed and personable than Brenda. She treated the other operator, Debbie, accordingly. Debbie was able to make personal phone calls, could take breaks when she wanted, and interacted with Andrea frequently throughout the day. Andrea really enjoyed her work and saw word processing as similar to a computer game. She was very talkative and informative during the interview. Andrea was not nearly as nervous about my observation as the other coordinators.[3]

Mary was the coordinator on the third floor and was midway between Brenda and Andrea in management style. She had been at the firm longer than any of the other operators. Mary had previously been a legal secretary but chose to go into computer operations when the company automated. There were two other operators on the third floor, Dawn and Francine. Dawn was a rover who was supposed to work at whatever center needed her, but she usually stayed on the third floor. She was bitter about having a coordinator over her because she felt she could do the work without a supervisor. Dawn felt as though the company "had taken the job and made it into something you cannot like, like this authority thing and then they run with it." Francine had just finished paralegal training but had not been able to find a position that paid as well as the word processing operator job. She was interested in learning more about the computer system and the company, but Mary was unwilling to train her in any job task which was not directly related to word processing. Both of these operators expressed some form of dissatisfaction with the authority structure.

Control in the centers took many forms. The attorneys who had direct involvement with the operators acted with a kind of indirect control. Even though their control was less direct than the coordinators' control, they still had the ultimate authority. The personnel director, Connie, held control over the coordinators. Dawn said that "Connie will come in, look at the work log and say 'oh you are not busy' or she will pass you in the hall and say the same thing." These kinds of comments were generally interpreted as threats. Connie kept records of productivity and had the authority to fire. However, the operators were controlled on a day-to-day basis by the coordinators.

The authority structure at Inland Attorneys was made problematic by general lack of communication between operators, coordinators, and the administration. This lack of communication created confusion surrounding personal and professional relationships. Because the coordinator position was perceived by most operators as a "boss" type of position and not simply as a coordinating position per se, the special skills involved were kept as protected secrets. This was exemplified by Mary who refused special training to Francine with the reprimand that such information was not part of her job description. Had Francine been able to acquire the skill, it could have functioned as an incentive (Blau and Scott 1962, Kanter 1977). The operators felt alienated from the personnel director. They did not understand her role and felt she was cold to them.

Work Process Most of the operators worked an eight hour day with one hour lunches and two 15 minute breaks. There were also times when they were required to work evenings and weekends. The starting salary was $1500/month. They were all well dressed and considered themselves professionals. The centers had very good equipment, such as a Jacquard computer system which was very powerful and not very user friendly. Each operator had a chair that cost $600.00.

The word processor operators were required to do word processing tasks, nothing more. The organization of the workflow was similar on all floors except that the type of the document changed. Most of the work came into the center via the attorneys or their secretaries. Then, the document was either put into a box or given directly to the coordinator. Whether the attorneys gave the document to the coordinator or put it in a box was usually determined by the importance and/or difficulty of the word processing task. If it was an important document or one that was complicated, the attorney would speak directly to the coordinator, if not, the document would just be put in a box. The coordinator would then distribute the document or do it herself. If there was too much work, someone from another floor would have to help out. Once the operator had the document, she would either begin a new computer file or revise an existing one. After completion, the file would be printed out and returned to the coordinator for distribution to the appropriate attorney.

Given the high standing of this law firm, the attorneys expected and worked with good word processing operators who were treated with a fair amount of respect compared to the other clerical settings I studied. They were also the highest paid. The operators had a collective identity which could be characterized as one of professionalism. Their orientation toward work was one of doing a good job and then going home and leaving it behind. The word processing operators' definition of work was that it was important to get the work done and to do it well. The decentralization of the structure that dictated three separate word processing centers made the distance between the user and the operator minimal. Although the coordinator was the liaison, the operators knew most of the attorneys and interaction as frequent.

The personality of the coordinator had some effect on the atmosphere of the center. There was an uneasy relationship between Andrea and the other two coordinators. Both of the other coordinators were uncomfortable with her informal style. Also, the hoarding of skills by Mary and the autocratic style of Brenda were problematic for the operators. However, it was the position of coordinator that created the most conflict for the operators. The operators did not understand why there had to be such a position and they wished that the position were eliminated. Most conflicts took place between levels of the authority structure. The operators did not want to be "babysat" by the coordinators.

Although the word processing staff was well trained, the authority structure took away from job satisfaction. There was not much autonomy for the operators and little room for innovation. The lack of communication between operators, coordinators, and administration created confusion. The authority relations and subsequent lack of communication between the levels of authority were the source of most complaints.

Western Bank

Western Bank is the second largest California based savings and loan company. The downtown branch is housed in an older skyscraper. This branch has an operations research department with forty full-time employees. The operations research department is comprised of a manager and two main departments. One department is Methods and Procedures which handles such matters as research, review, deposit, and investment support. The second department is Office Automation. This department is responsible for the word processing center, new office automation projects, and various support systems.

I obtained entry to the office automation department through Leah, the manager of Office Automation. I was referred to her by the same consultant who referred me to Inland Attorneys. Upon seeing the research proposal she gave me immediate permission to observe the central word processing center and to interview the operators at lunch or after work. I promised to write a final report discussing my findings at the conclusion of my study. She introduced me to the lead operator and showed me around the center. She was very happy to discuss office automation and my research project. I started observations a week after my meeting with Leah. I performed interviews with four operators, one supervisor, and three word processing analysts. I conducted seven hours of observation.

Authority Structure Western Bank has one central word processing center for its city wide operations. The center is located in its downtown San Diego branch. The word processing center takes care of the word processing needs of Western Bank for any document over five pages long and any document that required many personalized copies. The word processing center did everything from nation wide Visa rejections to departmental manuals. Western Bank employed four word processor operators and one supervisor. The word processing center had a Wang VS 20 system that was designed to process long, revisable, and repetitive documents as well as multiple letters and memos.

The authority structure was centralized. The manager of office automation reported to the manager of the operations research department who reported to the vice-president and so on up the line. Within the office automation division the tall authority structure continued. There were six people directly associated with the word processing center; three word processing operators, one lead operator, one supervisor, and the Office Automation manager.

Leah was responsible for the word processing center in addition to the implementation of other office automation projects, such as the Ad hoc Center and Distributed Word Processing Support. The Ad hoc Center did research and implementation for office automation systems, took care of linking personal computers to the main computers, and implemented teleconferencing. The Distributed Word Processing Support took care of the hardware for the decentralized word processing systems throughout the company.

Leah was the administrative boss to whom the lead operator and supervisor reported. She did not have much direct interaction with the operators unless it was in decisions regarding hiring, equipment, or conflicts in the workplace. I never did a formal interview with her but we spent quite a bit of time talking about office automation. Leah had a Master's degree in Business Administration and was quite bright. She was particularly interested in integrating some data processing into the word processing center.

Shawn was the supervisor of the word processing center and was hired just before I started collecting data the bank. Leah hired her because she had extensive data processing experience at another firm. Shawn's actual supervisory role of the word processing center was minimal. She spent most of her time learning the word processing system and doing research into the possibility of incorporating data processing into the word processing center. Shawn said that she interacted with the word processing operators every day, and she wanted them to feel that they could come to her at any time. However, she was perceived as a "straw boss" (Jaques 1978:211) by the operators. Shawn had the title of supervisor but was not perceived as having the necessary skills to fulfill the position. The supervisor she replaced had been close to the operators and did word processing herself. Shawn's pay was $1833.00/ month at an hourly rate.

The lead operator, Kim, was delegated as my direct liaison with the word processing center. She took care of most of the direct supervising of the operators and the work flow in the center. Kim's background was secretarial and she had a Bacelor's degree in political science. Her responsibilities included keeping account of the productivity of each of the operators through keystroke monitoring. These reports categorized the difficulty of the level of work and counted key strokes. This data went into monthly reports. Kim said that the reports were never really used unless there was some kind of problem with an employee. Kim's salary was $1310.00/month with a proposed raise to $1362.00/month. She knew more about the word processing system than anyone else in the company.

There were three other operators, Gwen, Carrie, and Joyce. Gwen was hired through a temporary agency. She was shy, bordering on neurotically quiet. She spent most of the time by herself and was very hard to talk to. My interview with her lasted about 10 minutes. I was not able to find out her salary because she worked through the temporary agency. Carrie was a bit more upbeat, but she was also quiet. She was 19 years old and wanted to go to the university to learn more about computers. Carrie was part of a new generation of operators who entered word processing without having a secretarial background. She earned $1160.00/month. The last operator, Joyce, had some prior secretarial experience and was the resident expert on a part of the system called indexing. She was also a quiet person and did not make any waves. Her salary was $1210.00/month. These operators had responsibility for typing and revising documents for users as well as file maintenance.

There were bureaucratic and technological controls over the operators at Western Bank. The tall authority structure harnessed the employees' autonomy. Kim said that

> I don't agree with this style (of management) and how they run things. Everybody has different management styles and I don't

agree with this one. I have a hard time—there's enough aggravation and frustration because we're in a time-pressure situation. There's a lot of pressure in there, okay. And there's, you know, plus the workload. This is a management style of looking over the shoulder, they are more concerned with productivity than with the way people feel. That's a very important aspect of your job, how you're treated, because we do work as a team in there, when one person is upset, it affects everybody."

This authority structure led to conflict between levels of authority similar to the conflicts at Inland Attorneys. Leah, the manager, had bureaucratic control over the Office Automation staff but hardly exercised it. The center ran without her direct supervision. The supervisor was not in the same room with the operators and did not exercise direct control either. The lead operator was the person who dictated and distributed work, although she was not a formal supervisor.

The authority relations were centralized and the operators felt alienated from the rest of the firm. The operators wanted to feel liked by the rest of the firm but felt they were not. Carrie and Kim complained that they would pass Leah in the street and she would not speak to them. This created hostility. Leah said that she was too busy talking to other managers to say hello. Leah felt that the operators should understand that she was involved in other things and not let their feelings get in the way of professionalism.

Another example of the operators' alienation was that they were not informed about the description and duties of Shawn, the supervisor, who was making decisions about the center. Nor had they been given any part in those decisions. Because Shawn was hired to do research and development with data processing and did not know the word processing system, there was confusion between the role of the supervisor and that of the lead operator. Because Shawn was paid considerably more, this led to resentment. As a result of lack of communication and understanding, the operators felt isolated from management and the supervisor.

On the other hand, management did not understand the operators. Very often management assumed that operators were only at the level of a clerk typist. Management lacked a clear understanding of the job duties as well as the overall running of the computer system. The amount of knowledge required to run the system was underestimated by some in management positions. In my discussions with the word processing management, there were pervasive feelings of separation from the operators. Leah talked of the word processing center with pride and would raise issues about the kind of machines they were using. She and other Office Automation staff would focus on the equipment and not the people. The operators would talk more about the people. The lack of understanding among the levels of the authority structure led to poor relations.

Work Process The operators worked staggered hours at Western Bank. Some would come in at 6:30 a.m. and some would not come in until 9:00 a.m. The center was open from 8:30 a.m. until 6:00 p.m. for the users. The operators had a half hour for lunch and no scheduled break times. Kim, the lead operator, worked a four day week with 10 hour days.

The organization of the word processing center at Western Bank was quite a bit different from the one at Inland Attorneys because it was centralized and had hundreds of users. There was very little interaction between the users and the operators. Most of the time the operators did not know the users. The users would drop off the work in a basket at the front desk or send it through the company mail. There was almost always a continuous flow of work which would first go to the lead operator and then be distributed to the other operators. After completion, the document would be sent through the mail or be picked up by the user.

The users would fill out a form and put it in a basket to be distributed by the lead operator. The form asked questions about the document such as when the document was needed and if there was an earlier version. This information was important because it helped prevent a document from being late or an operator from entering a document twice. The work had to be submitted in typed or legibly written form, exactly the way the user wanted it, including formatting and spacing requirements. The operator was responsible for getting the document to the user by company mail or a phone call.

One of the operators, Carrie, was responsible for the weekly "backing-up and restoring" of the computer files. This is essentially copying the information coded on the computer disks onto computer tape so that there would always be a copy of vital information should the main computer "crash" (i.e., inadvertently erase information). Another operator, Joyce, was responsible for making sure the other operators had the best and most efficient programming. The temporary, Gwen, received odd jobs. These operators were required to do word processing related activities and nothing more. Yet, the women all expressed pride in having these specialties on the computer. Carrie said "I think my favorite aspect of the job is backup because it gives me the feeling of—not having authority or power but knowing—knowing how to do it. It makes me feel good, knowing that I can do it. I'm not kidding." The three permanent employees were interested and enjoyed working with computers and were very good operators.

The work process design at Western Bank was centralized and productivity was monitored. There was little interaction between the operators and the users. This gave the operators a sense of alienation and isolation. However, the actual computer operations gave the operators a sense of pride. Given the volume of the demand placed on the word processing center and Western Bank, there was never much spare time for the operators. The centralized nature of the center made interaction impossible. In virtually all the interviews the operators felt isolated from the rest of the company. The perception was exemplified by Carrie when she said "I don't know what she (the new supervisor) does, why don't you find out and tell me." Not only did the operators not understand management, but management did not understand the operators.

The workplace was strained by a lack of communication between the operators and those above them in the authority structure. The operators wanted to be liked by Leah, Shawn, and the rest of the Office Automation staff, but there was tension regarding the employee's perception of the supervisor and the management's perception. The previous supervisor had been close to the operators, the new one was seldom around, and when she was she "watched over their shoulder." The operators' collective identity was primarily held together by their dislike and distrust for their supervisor and manager. The operators' definition of work was based on an ongoing attempt to not get too upset at their situation or with their superiors. However, the operators really liked to work with computers, and the relationships among the operators were strong.

Beach Attorneys

Beach Attorneys is a law firm that is located in a very prestigious beach community in San Diego. The offices are in an impressive, brass trimmed bank building surrounded by restaurants, boutiques selling expensive clothing, hotels, and investment companies. Beach Attorneys provided a whole range of legal services including both corporate and criminal law. The firm is comprised of six partners, seven associates, and approximately 27 support staff. The support staff included legal secretaries, law clerks, paralegals, a receptionist, three word processing operators, an accounts receivable clerk, an accounts payable clerk, a legal administrator, and her assistant. They have one computer room that houses the two account receivable personnel and three word processing operators.

I obtained entry to this firm by working as a data entry operator. They advertised for a temporary full-time operator to replace someone going on medical leave. I called, got an interview, and was hired for the job without previous data entry experience. I wore a suit to the interview which turned out to be appropriate because the administrative assistant, Pat, said that they had wealthy clients and dress was important. I was hired to work for six weeks from the end of July through the middle of September 1985. As it actually happened, I worked full time for four months and on and off for another two months. The duties I was initially hired to do were data entry and to cover for the receptionist while she was gone on break or to lunch. However, I also ended up doing such tasks as stocking the kitchen, cleaning the supply room, running errands, and making car payments for my supervisor.

The personnel, both attorneys and computer operators, knew that I was doing a study of word processing centers and that I was a graduate student in sociology. However, I did not make explicit the fact that I was taking field notes on my experiences at Beach Attorneys. I conducted two tape-recorded interviews and six months of participant observation.

Authority Structure The computer room personnel were responsible for all of the word processing needs and most of the data processing needs of the firm. Any document over three pages long, and all the accounts receivable for the attorneys were processed by the center. All of the people in the computer room were accountable to Mary and Pat, the head administrator and her assistant, respectively.

I was at this setting much longer and in a totally different capacity than I was at Inland Attorneys or Western Bank. Therefore, I obtained considerably different more in depth kinds of data. I was aware of the functions and the organization of the company, not just the computer operations. However, I never learned anyone's salary because they were confidential, but I assumed their salaries were in the high range because the firm was prestigious. I earned $7.00/hour as a data entry operator.

The authority structure was relatively decentralized because of the size of the firm. The computer operators reported to the legal administrator who reported to the attorneys. There were two major categories of attorneys—the partners, and the associates. The partners owned the firm while the associates were salaried. For the most part, all of the attorneys treated the support staff, particularly the clerical workers, with little respect. The attorneys also had a tendency to give orders to people, rather than ask them when they wanted something. The authority structure and degree of supervision were very autocratic in this small firm.

Mary was the legal administrator for the firm. She was in charge of everything and everyone, except for the legal cases and the legal staff. This meant that she furnished the office, hired the staff, and had responsibility for

Historical, Ethnographic, and Comparative Accounts 57

supplies and equipment. She would also do whatever personal errands were demanded by the partners. The firm revolved around her. Mary had been with the firm for the twelve years it has been operating and had no formal higher education. I assumed that she was paid very well. Mary was on a six month leave while I was at the firm. Although she wasn't there, she was referred to frequently. Most people had negative things to say about her, with the exception of Pat her administrative assistant. Mary was the "right hand person" of the attorneys and was autocratic. The rules and regulations created by her were very rigid, thus setting the stage for uneasy relations, not only between the attorneys and the staff, but among staff members as well.

Pat was Mary's administrative assistant. She was in charge while Mary was on leave. She could call Mary if she encountered any problems. It was a very difficult and stressful time for Pat; she had a lot of responsibility. She did manage to make it through until Mary returned; however, she was quite ragged at times. Pat had a much more flexible supervisory style than Mary and was well liked by most of the staff. However as time went on she, too, seemed to become more autocratic.

Sharon was in charge of accounts receivable and supervised the data entry operator (me). She did not officially have a title but her responsibilities were quite extensive. Sharon was in charge of all the billings to clients in addition to reports to the partners' monthly meetings in which she would calculate how much was owed to the firm and how much the firm had been paid. She also had to keep the computers up, the programs working efficiently, and store all the information. She was going to school part-time at the local state college.

I was the data entry operator and I entered numbers and bits of text which related to the logging of the attorneys' time. This data was used to bill the clients. The clients were billed by the tenth of the hour. The attorneys were paid up to $220.00/hour. Time sheets were filled out and entered into the computer system throughout the month. The time sheets would then be tabulated and the machine would print out bills. These bills would be sent to the attorneys or their secretaries to be proofread and sent to the clients. The most difficult part of the data entry job was making sure there were no mistakes on the bills. This meant that the hours, dollars, computer totals, and handwritten totals had to correctly add up. Additionally, I was expected to stock the kitchen and supply room, run errands, and relieve the receptionist for breaks and lunch.

On the word processing side of the computer room there were three operators. Randy was the supervisor. Barb and Rosemarie were operators. Randy had a Bachelor's degree in English and Barb had a Bachelor's degree and some units toward a Master's degree in business administration. I did not know Rosemarie very well because she worked a swing shift. The operators were responsible for all the word processing needs of the firm which the legal secretaries could not or did not want to do.

Randy spoke his mind frequently and was not intimidated by the attorneys or the administrative staff. He ended up getting most of his supervisory duties taken away from him due to "personality problems." Randy typed well over 100 words per minute. He had been a word processing operator at a large defense contractor before Beach Attorneys. His hope was to work himself into a managerial position from word processing. But, as time went by, Randy started to learn that promotion out of clerical occupations was difficult.

Barb was from the Philippines. Barb worked for Randy. She was a very good employee. She never avoided her responsibilities, was on time, deferred to the bosses, and liked word processing. Her previous job had been as a manager for an accounting office. She quit because there was too much stress and pressure. She too could also type over 100 words per minute. However, Barb also had some problems at Beach Attorneys. Her trouble seemed to be linked to the autocratic style of the firm. She told me about a time she took off work—which she did not do very often. Her husband was in the service and he had come home on leave unexpectedly; she took time off work then. Shortly after that she took a couple days off because she was sick. This is what she said

> I mean, you would think that the legal administrator would be a little human or something. Nope. They tell me `This is work.

You've got responsibilities. You have to set aside your personal feelings and we don't care what that is.' And so, anyway, I was sick and then I was sick again last week or two weeks ago. I have been sick. Pat (the legal administrator) came to me and said 'you now that you're going to be out, you know you gotta' let us know in advance.' As if I know when I'm going to be sick! You know, I cannot plan on being sick. She said, okay. So I am afraid that this is, again, being put on my records.

This treatment of Barb is a good example of the type of authority relations at Beach Attorneys.

The policy and procedures for the support staff were autocratic. The staff was not allowed to call the attorneys by their first name during business hours. The staff, particularly the receptionist, could not eat, drink, or read at their stations. There was a policy for the receptionist to record people for being ten minutes late—their pay would be docked. These kinds of rules made the working situation uncomfortable. The control in the computer center was bureaucratic, although direct supervision was nominal when Mary was on leave and Pat was in charge. Mary controlled the clerical staff through intimidation and the threat that they might lose their job. Pat utilized a less strict style and was liked by most of the staff.

The authority structure at Beach Attorneys created the most problems for the workers. This was particularly apparent in the treatment of the support staff by Pat and the attorneys. On several occasions I witnessed a staff member crying because of an attorney's treatment of her. It happened to me when I accidently cut a client off the phone line and an attorney became hostile. The humiliation took its toll. There had to be a great amount of deference to the attorneys. Other than the attorneys' harsh manners, the staff complained about unpleasant tasks they had to do. Because it was a small company, everyone was expected to pitch in if necessary. This, too, would cause resentment because the activities were usually outside of the staff's job descriptions.

Work Process Most of the support staff at Beach Attorneys worked from 9:00 a.m. to 5:00 p.m. They had one hour lunches and no formal break times. All of the firm's computer operations took place in one room. There was one hard drive for data processing and one hard drive for word processing. It was physically impossible for the operators to look at one another while they were working. The data processing and word processing departments were separated by a divider over which they could not see. Both the data and word processing departments' printers were suited to their particular needs. The divided room was glass on two sides so that if the blinds were open, the operators could be seen by the attorneys and staff as they walked by. There was indirect lighting and a pleasant, comfortable, and soft atmosphere. There were no outside windows.

The accounts receivable section processed time sheets from each attorney. On those sheets the attorneys recorded the amount of time they spent and what they did with each client. These time sheets were used to bill the clients. The data entry clerk entered the data into the computer in the form of numbers and words. The time sheets would be placed in a box. The data entry clerk had almost no interaction with the attorneys or secretaries. The exception to this was when there was a question about information on a time card. The accounts receivable section was very busy at the end of the month because there was pressure to get out that month's bills. At the beginning of the month the data entry position was very slow.

The word processing operators had responsibilities to specific attorneys. The attorneys or their secretaries would drop off a document, talk to the word processing operator about the document, and come to some kind of decision about a time frame. The word processing operators would have to keep up on all her or his attorney's work. The work process design at Beach Attorneys was less routinized than Western Bank in the sense that the staff would do varied activities depending on the amount of work in their department. However, having to do work outside one's job description caused resentment. The data entry position was the least skilled and most routinized of all the positions. There were minimum computer skills required of the data entry clerk.

Because the size of the firm was small, there was some interaction between the users and the operators. However, the strict authority relations, policies, and procedures created a strained atmosphere at Beach Attorneys. The definition of work was dominated by the negative group dynamics between different levels of authority. During my employment one of the primary concerns was the return of Mary. Her return would mean the end of the informal atmosphere that Pat allowed. There was also not much autonomy or sense of freedom as a result of the glass walls that enclosed the computer room. For the most part, the ties between employees were weak. The uneasy relationship between the bosses and the clerks, and between the clerks, made job satisfaction low.

Sporting Goods International

Sporting Goods International is a multinational corporation that produces and distributes various sporting goods including tennis shoes, racquets, racquet ball equipment, and volleyball shoes. The company is owned by Korean business people and most of the manufacturing takes place in Korea. Sporting Goods International is housed in a fast growing inland commercial center and sells and distributes the company's products, mostly in the United States. The firm does over a million dollars of sales each month. Its headquarters are neither extravagant or dumpy, and it houses the president, vice-president, head consultant, customer service, accounts receivable, and accounts payable departments.

I obtained entry to this firm through Conrad, the head consultant.[4] He was looking for someone to develop a sociology of work course for him. He hired me to develop the course.[5] During the beginning part of this project I realized that there was a computer center, so I asked and received permission to interview the support staff. The company was in the process of converting from an IBM 34 to an IBM 36 computer. It took me about two months to interview the whole staff. I performed ten interviews.

Authority Structure All data entry took place in one computer room which was responsible for billing. Because the firm was in sales, billing was an arduous task. There were three computer operators. Two of the operators were data entry clerks and the other was the supervisor and resident computer whiz. There was no word processing center. All word processing needs were taken care of by individual secretaries. The authority structure at Sporting Goods International was quite different from the other three firms. There were very few titles among the support staff and they were urged to talk about problems and solutions with their bosses or at the company meetings. Overall, the authority structure was extremely decentralized.

The head consultant, Conrad, had most of the power in this company. It was quite obvious by the deference given to him that he was the boss. The president of Sporting Goods International seemed to be a figure head who took care of the day to day operations but was under the supervision of Conrad. The officers of the company were all part of the consulting firm that Conrad operated. The consulting firm was hired by the owners to run the firm.

There were ways that Conrad tried to give the office a comfortable and nonhierarchical atmosphere. He had the office designed as an open space. Only the executives had offices. Conrad's and the president's offices had doors with no windows. The other executives had offices but there were large windows facing the rest of support staff. There were no cubicles or offices for the support staff. They were all housed in the center of a large room. The executives' offices surrounded that room. Conrad explained that the idea behind the open office was that there would be more of a family feeling. However, I had the feeling that it also worked to keep an eye on the support staff.

Conrad wanted the employees to feel that he was approachable, but his intimidating style made that nearly impossible. Another way he attempted to make the office nonhierarchical was by not using titles for the support staff. He argued that there was less separation between the management and the workers. But a latent function was that it was possible to pay people less. For example, Jennifer who ran the computer room essentially performed data processing managerial duties, but since she was not called manager she could be paid less than a managerial salary. Conrad's techniques for creating a family feeling could also be used as mechanisms for control.

Pam was Conrad's secretary, wife, and office manager. She was the liaison between Conrad and the rest of the staff. She seemed quite subservient to Conrad and would not make decisions independent of him. However, the support staff knew she held a pivotal position and they treated her with respect. Actually, they seemed to really like her. I got the feeling she was Conrad's "informant." She did not want to be interviewed.

The acting office manager, Gail, was responsible for the support staff when Pam was not there. That staff included three customer service representatives, three collections department personnel, and three workers in the computer room. In addition there were three secretaries under her. She was very busy and made $24,000/year. Gail was one of the most relaxed people I interviewed at Sporting Goods International. She was friendly, articulate, and quite open to talking about any issue. Gail told me that she thought the best way to manage people was to expect that everyone was an adult and would act as an adult. The only thing she disliked about her job was that she was continuously interrupted. She said she liked everything she did.

Jennifer was the head computer operator. Her duties included supervision of the two computer operators and the running and maintenance of the computer. She knew much more than anyone else about the computer and peripheral equipment. Jennifer was the only person who was able to get information or reports out of the machine. All of the sales and shipping information were in the computer and it was her responsibility to manipulate the data and distribute it in readable form. She held a key position but had relatively low status compared to the managerial staff. Jennifer was nervous and animated, but self-confident during the interview. She held some animosity toward Conrad regarding how she was treated. Her knowledge and contribution to the company were extensive.

The two operators, Jean and Sara, worked for Jennifer, respected her, and seemed to get along well. Jean had been with the firm for one-and-a-half years. She made $6.32/hour. Jean was resentful of her pay rate and blamed Conrad for it. Jean was concerned that Sara, who was recently hired, was making more money than she was. She said that "she (Jennifer) has a lot of control over us, but Jennifer and I get along fine. I know my job so she pretty much leaves me alone. I get along with everyone else well, except Conrad. That's pretty shaky." Jean liked Jennifer but felt disliked by Conrad. Sara had a bubbly personality and was anxious to learn more about the company. She really liked her job and the people with whom she worked. However, Sara did not like being taken for granted by others. Sara had a Bachelor's degree in Fine Arts. She had been with the firm for two and a half months, was happy with her position, and hoped to be working her way up. She made $900.00/month.

The type of control at Sporting Goods International was bureaucratic and charismatic. The office manager had authority over the support staff, but she did not directly exercise that authority, nor did she directly supervise them. Most of the control was charismatic—coming from Conrad. The authority structure at Sporting Goods International was decentralized and flat. The network of relations was relaxed. There was socializing between the operators and the rest of the company. The open office environment was developed to make the work place more family like but also seemed to make people feel on stage.

Work Process The employees at Sporting Goods International worked from 7:00 a.m. to 4:30 p.m. Monday through Thursday, and on Fridays they worked from 7:00 a.m. to 1:00 p.m. They saw the short workday on Friday as a real benefit. There was a regular monthly meeting which all of the employees attended. During these meetings they talked about business profits, organizational issues, and projected sales. There was also time for the employees to talk about any problems they might be having.[6]

The organization of the work process was quite different from that at any other firm I studied. This is related to the fact that it is a completely different kind of business with different needs and therefore different kinds of organization. The computer operators entered orders that came from the shipping operations and customer orders departments. The first priority of the computer room was to enter the customers' orders. Jennifer would then run the "packing slips." These slips are packing and mailing forms which were attached to the actual shipments. That is essentially what the computer room did except for running management reports at the end of the month.

There was a fair amount of communication between the users of the center and the operators. Even the shipping clerks knew the operators. The orders came in from all areas of the company, including secretaries, the order department, sales representatives in the field, and the shipping area. There was a sense of cohesiveness and understanding about the different types of jobs. The support staff was small and, on the whole, there was a relatively high amount of interaction between the ranks. I was as likely to see a computer operator walking with an executive as with a secretary. There was a team spirit and most of the employees knew a lot about the sporting goods equipment that was being sold. However, although only a couple of people mentioned it, I felt like people were scared of Conrad and most of the employees' problems were implicitly associated with Conrad's rough style. Nonetheless, the workers were generally happy and not too alienated.

Comparative Account of Organizational Designs

The companies' profit-oriented goals determined management's perception of the support staff. These firms saw clerical workers as "overhead." The result of this perception was that authority structures and work processes were directed toward keeping the employees under control and thereby efficiently producing services and/or commodities. However, differing degrees of centralization and rationalization existed. The organizational designs of the firms in my study effected workers' perceptions and activities and were related to worker satisfaction. The authority structures were all hierarchical, but they varied in the level of centralization and types of control. The work processes varied according to level of rationalization and technology. Authority structures were the most influential in determining worker satisfaction. In the four cases examined here, less centralization and less rationalization were related to higher worker satisfaction.

Authority Structures and Centralization

The primary source of clerical workers' complaints was the authority structure. The firms in this study had varying degrees of centralized authority structures. The more centralized the structure the more complaints. Decentralized authority relations led to more humanistic social relations characterized by an increase in communication and worker satisfaction, and a decrease in alienation.

Western Bank had the most centralized and tall authority structure. The organizational chart contained many levels of authority. There were strict formal rules and not much communication between the levels of authority. Their word processing center served the needs for the San Diego region but there were only four operators. The stress and subsequent dissatisfaction and alienation seemed more pronounced than at the other settings. The management style was autocratic. This structure negatively affected the word processing center and the operators. Management made decisions regarding operators, with little knowledge of how the operators actually did their jobs or how they felt about the changes. This power differential left the operators with a feeling of inability to deal with or improve their situation.

There were explicit and often talked about conflicts at Western Bank. One conflict centered around the new supervisor. The operators liked the old supervisor who was moved to another department. The new supervisor did not relate to the operators and did not know the word processing system. The operators also resented Leah, the manager, for not speaking to them in passing. Yet, the nature of Leah's job made it impossible for her to relate to the staff in any personal way. Poor relations between management and operators plagued the center. There was a general lack of communication and understanding of work demands at all levels. The workers were not given access to decision-making in their work place. The centralized authority structure was clearly related to worker dissatisfaction.

A good comparison is with Sporting Goods International where the authority structure was significantly less pronounced and where there was more access to company information and decision-making. The company's rules and regulations were more informal. The communication between levels of the hierarchy was much greater. This was exemplified by monthly meetings during which the employees heard about profits and were given a chance to air their grievances. The clerical workers had more interaction with employees other than their peers. They were interested in the company and wanted to know more about it. They did not display nearly the level of complaints as did the operators at Western Bank.

Inland Attorneys and Beach Attorneys both had roughly similar levels of centralization within their authority structure. There was a supervisory position and a management position between the computer operators and the users. At Inland Attorneys the source of greatest operator complaints was the coordinator (supervisory) position. The operators saw no need for such a position and resented being watched over by a peer. Beach Attorneys had similar levels of authority structure, but the policies and procedures governing the work was more autocratic than at Inland Attorneys. This led to employee resentment and lower satisfaction. There was some operator input on decision-making and some low level communication between the users and the operators. The employees seemed to have greater job satisfaction than those at Western Bank but less than the workers at Sporting Goods International.

In sum, centralized and autocratic authority structures were the primary source of job dissatisfaction. Centralized authority structures and strict control within the firms created employee dissatisfaction. The authority structures were apparent in the differences in level of direct supervision, avenues for communication, and access to company decision-making. Strict control was counterproductive because it caused alienation and lack of communication. Within organizations with similar authority structures there was variability of operators' satisfaction depending on the formality of policies and procedures. Authority structures set the parameters for control and worker satisfaction, but were not all-determining.

Work Process and Rationalization

The computer rooms included in this study had differing work processes but similar workflows. The users would bring the work in, the supervisor would distribute the work, and then the finished document would go back out of the center. However, within this broad framework there were variations depending on the type of business and authority structure. The operators were generally happier in situations in which they had more autonomy, a broader range of work responsibilities, and were not closely monitored.

Rationalization was most apparent with the tabulation of key strokes and a continuous workflow. Western Bank was the only office operating under these conditions. Although the lead operator said that the tabulations were not used, the operators knew that their productivity was being monitored. Western Bank also had a strict process for users of the word processing center. They had to fill out work-order cards which the operators received through company mail. This process negated most reasons for the users to talk to the operators. In turn, this created a sense of isolation for the operators. On the other hand, the operators at Western Bank were given a wider range of duties to perform than just entering words. Certain specialties were assigned to the operators which gave them a sense of pride. For example, it was Carrie's responsibility to do a weekly back-up and restore of all the computer files. This responsibility made her feel like an important part of the company. Kim, the lead operator, had numerous supervisory responsibilities which relieved her from continuous typing. She, too, enjoyed having a specialization. However, even though they had a sense of pride regarding their particular areas, there were more complaints here than at the other settings.

Inland Attorneys also kept records of productivity, but these records were not computerized, they were handwritten lists of documents done by specific operators. The operators were under the supervision of the coordinators. The operators and coordinator were housed in the same room, therefore, the supervision over the operators was direct. The coordinators did not have anyone directly watching over them. The coordinators determined who was to get what job and were responsible for keeping the computers and printers functioning smoothly. The coordinators had a wider range of responsibilities, and some degree of control over the work process. They did not complain as often as the operators. In other words, the coordinators had more autonomy, wider range of duties, and experienced more satisfaction.

Sporting Goods International had a very different type of work process but it was quite routinized. Because it was a data processing center, two of the operators did nothing but enter numbers and occasionally file. They were off in a room by themselves and had the opportunity to have quite a bit of private interaction, mostly joking and fooling around. There was flexibility in the time table for job completion. The operators could put off certain activities in order to work on learning more about the new system which was being installed. So, even though they did not have day to day work that required a great deal of knowledge, they were relatively satisfied with the flexibility of the work process and the interaction that took place.

At Beach Attorneys the word processing operators' work process required more knowledge and skill than at Sporting Goods International. The operators were in close proximity to the attorneys and the documents they processed were sometimes very technical and difficult. Becky, a word processing operator, talked about times when certain attorneys would ask her opinion about how something was written or spelled. She really enjoyed being asked for her opinion on the documents. She liked working with attorneys who regarded her as a knowledgeable person. In this case, the knowledge and difficulty of the duties was an important source of satisfaction.

Conclusion

The use of strict control over the work process by management was dysfunctional in all cases. It actually worked against the productivity it was designed to increase. The amount of control over the work process that an employee was given was related to job satisfaction—the more control workers had the more they were satisfied. The more autonomy and responsibilities an operator had, the more likely she or he would be satisfied with his or her job.

The effect of rigid control in work process design was the same—workers felt alienated and dissatisfied.

The authority structures and work processes of automated offices were instrumental in determining workers' perspectives and activities. The companies that sacrificed classical authority structures and strict control over the work process had more satisfied workers. In all of the firms the main sources of conflict were the authority relations and lack of worker control over the work process. Thus, from a managerial point of view, the greatest productivity would be achieved by developing organizational designs with flexible authority relations and work processes determined by the workers. Worker satisfaction was related to wide range of responsibilities, short authority structures, and communication between levels of authority structures. Small range of responsibilities, tall authority structure, and lack of interaction between workers created worker dissatisfaction.

Notes

1   During the data collection phases of this research project I employed various techniques to ensure anonymity. Any marks of identity were changed or deleted in both the transcripts and the presentation of the data. I used pseudonyms for any possible marks of identity relating to the participants and the workplace. Therefore, the names of the participants and the businesses have been changed.

2   The promise to submit a final report was made because I felt it would assure entry; they got a consultant's report and I got the interviews. There is a flaw here because it took me out of the role of researcher and into a consultant's role, i.e., management's helper. Some of the operators initially thought that I was doing a time-motion study. However, I feel that the manner in which I conducted the interviews mitigated most of the negative effects. For example, during the interviews I would stress that I was interested in their story, how they felt, and not in their productivity. In this way, I tried to show the clerical workers that I was on their side.

3   There had been a time-motion study done just previous to my data collection. I think that Brenda and Mary thought I was involved in that previous study or was doing something similar to a time-motion study. They were uptight.

4   Conrad had originally been the president but chose to give up the title and create a consulting firm to run the business. The new president and other executives had worked for Conrad before and were under his control.

5   Conrad had taught sociology at the local state college before he went into business. He was planning on vacationing for one semester in the city where his daughter was attending an university. He was hired to teach a sociology of work course at that university. He needed some help catching up on the literature.

6   During the one monthly business meeting I attended Conrad did most of the talking with the other executives speaking very little. The support staff

only spoke when spoken to and then just briefly. I think the executives wanted to hear from everyone, but the support staff did not feel comfortable enough to say much.

4

OFFICE PRACTICES

Specialization, rationalization, and sexism comprise some important elements of the macro-social context and of the organizational designs of automated offices. To get a better view of what is going on in the automated office, we need to look at specialization, rationalization, and sexism as taking place on many levels, not just the macro-social and organizational design levels. The informal organization of the automated office is another level of social reality which unmistakenly influences the clerical workers' work world. By informal organization I mean interactional and behavioral patterns that comprise the office. The clerical workers' informal organization has as much impact on their work world as do the macro-social context and organizational design. Macro-social context and organizational design are better understood through an in depth analysis of behaviors and perceptions of clerical workers.

In order to talk about the clerical workers' informal organization I analyzed their everyday life practices in the office. I found two distinct types of office practices—official and unofficial. I use the terms official and unofficial because I found that some behaviors were reflexive of organizational design, and other behaviors were strategies to get around organizational design. I refer to those behaviors which were reflexive of organizational design as official, and those behaviors that circumvented organizational design as unofficial. Figure 1 illustrates the relationship between the clerical workers, the organizational structure, and office practices. Clerical workers' subjectivities, including their perceptions, values, and needs, combined with organizational structures to create office practices. Unofficial office practices were more closely related to clerical workers' subjectivities and official office practices were more closely related to organizational design.

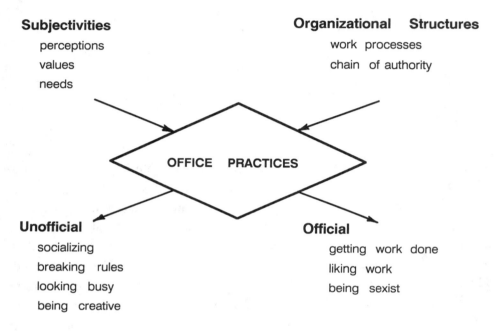

Figure 1. Office Practices

Official Office Practices

Official office practices were behaviors that functioned to create and maintain specialization, rationalization, and sexism. These behaviors depicted an acceptance of the organizational design and acted to uphold it. The clerical workers' perceptions and activities did not even hint at the idea that there may have been something wrong with the organizational design under which they worked. Of course, there were grumblings about supervisors and bosses, but there were no criticisms of the overall legitimacy of that structure. The authority structure was without fault, so the people occupying the positions in that structure were the problem. Personalities were blamed for any problems.

A traditional macro-social explanation of the type of perception explained above is that of alienation. Such explanations suggest that people are alienated when they become estranged from the products and processes of their work and also their fellow employees and, ultimately, themselves (Marx 1867, Blauner 1964, Erikson 1986). It is argued that workers accept subordinate positions because they do not perceive their position as oppressed. The workers believe that their place in society is proper, and therefore, the oppressed position becomes their place. Charlotte Wolf (1986) analyzes subordination from the level of the oppressed and argues that people submit to oppression because they reflexively legitimize it. She says that "the legitimation of oppression is one of the key issues for subordinate people, and that the study of the creation and processes of legitimation among such groups will shed light on the age-old questions of human obedience and of resistance to oppressive rule" (1986:217-218). She uses a comparative-historical method, analyzes three different oppressed groups, and develops a notion of "reflexive legitimation" to explain why workers submit to and accept subordination. Reflexive legitimation is a process in which workers rationalize their own subordination.[1]

However, individual behaviors that constitute the acceptance or legitimation of an oppressed position are not included in the analyses by macro-social sociology or Wolf. Yes, the workers are alienated and, yes, they accept the social structure, but how does that really happen? What routine behaviors and attitudes create and maintain bureaucratic control of the clerical worker?

There were three general office practices which demonstrated an acceptance of organizational design. The practices were to

1) get the work done,
2) like the work and the company, and
3) perceive themselves and other women as inferior to men.

These official office practices were a way that the clerical workers legitimated and accepted organizational design.

Getting the work done was the most important official office practice that kept the offices functioning and maintain organizational design. The work flow, or how much work was coming into the centers, set the parameters for the pace of typing. If there was an overload of work there was never a question that someone would have to work overtime. There were times when it seemed unfair that the clerks had to work overtime but they were resigned to it. The following examples illustrate this attitude and acceptance.

--The operators at Inland Attorneys had been planning for two weeks to go to one of the new hotels in the valley for a "coloring" demonstration during lunch.[2] But an attorney (the top partner's son) came in with a 100 page document and needed it immediately. Two of the operators had to stay through lunch and have no breaks until it was done.

--When Sporting Goods International was converting to a new computer system, the employees were expected to work "comp" time to learn the new system. They had to come in, they did not have a choice. Comp time is a system in which the workers worked extra hours and were later able to take comparable time off. They did not get paid money for comp time, only hours.

Anytime there was work to do, it was given first priority. The anxiety level was high when there was a lot of work in the computer centers and everyone contributed toward completion of the job.

The second official office practice was to *like the work and the company*. Most negative complaints about the company were not tolerated by the support staff. It was acceptable from time to time to articulate a sense of boredom or dislike for certain aspects of one's job, but a sense of an overall dislike meant trouble for the tenure of that person. For example Randy, a word processing supervisor at Beach Attorneys, was a constant complainer—he was demoted. Liking their work was an important part of the clerical workers' ideology and those who did not like their job often experienced a sense of isolation. The explicit acceptance of one's position and company was pervasive in my study. What I found is that if employees did not like their work and said so, tension ensued. Most employees said they liked their work. This kept the centers running.

The importance of the practice to like the work and the company was exemplified in the notion that "the boss is always right" because he or she is part of the company. This is mostly done on the front stage because the boss is often ridiculed backstage for her or his ill-informed or eccentric notions and needs. On the other hand, it is unacceptable to do anything but obey the boss. Arguing with the boss was a problematic and dangerous thing to do if one was interested in keeping her or his job. In conjunction with this, there was a practice to "keep fellow employees in line"—to follow the rules. For example, during my participant observation I was once reminded by a secretary that I was not to use the phone for personal calls. This person had no authority over me, she was just being a "company woman." Liking the work and the company were strong cultural mandates and constituted official office practices, which clearly maintained existing authority relationships.

The third practice that reinforced the status quo was the clerks' *perceptions of women as inferior to men*. Again, by official office practices I mean those behaviors which support organizational design. Sexism is an insidious and powerful form of control over people. Sexism was quite pervasive and the clerks had internalized sexist ideas that operated in insidious ways. For example, some employees viewed women as "catty." They saw women as being harder to work for, less objective, and more spiteful than men. They also neglected and/or denied their own education, interactional, and computer skills. This self-and other-directed sexism served to maintain and reinforce negative attitudes about women and work (Baker 1987).

Self-directed sexism maintained organizational design and was, in part, constituted by a negation of one's own education and skills. The clerical workers did not define their organizational, interactional, or computer literacy skills as important in their repertoire of work skills. Additionally, any higher education was almost completely irrelevant to them. Denying their skills and education mitigated the effects of being highly skilled and educated in dead-end positions and reflected a sexist view of women.

Other-directed sexism also helped to maintain organizational design. Andrea, the coordinator from the second floor of Inland Attorneys, had this to say about women and men.

> Women tend to be more controlled by their emotions sometimes than their minds, or their feelings than their minds. Men seem to be able to remain more objective; they may feel the same thing, but they don't let it affect their work. They can still handle situations in a mature, professional, and business-like manner. Women tend to go behind each others' backs rather than to your face. These things can disrupt more than the machines.

Beryl, a customer service representative from Sporting Goods International, said this.

> Working with so many women, I find it trying at times. Women can be catty and all women are moody. As far as when they have their time-of-the-month and you have to deal with all these people that are on the rag this week. It is always somebody who's having a hard day.

Jennifer, the head computer operator at Sporting Goods International, stated the following.

> If you don't get along with somebody, you take it personally and I think maybe that has to do more with women than men. I used to work with men and I didn't have too much of a problem, but if women don't like one thing about you they don't like the whole person. They don't separate work and personalities.

Self-and other-directed sexist practices stem from a form of internalized sexism and stand as examples of the power of macro-social forces. Both men and women internalize sexist notions. The women in my study were angry about their position but they had no voice for it. In some cases they were aware of their secondary position, but what could they do about it? They had also internalized an ideology which hindered the development of strategies to mitigate the oppression. I am not trying to blame the victim; I don't see it as their fault. The ideology and practice of gender inequality is very powerful. Sexism is a dialectical relationship that takes two to create and maintain.

In brief, it was accepted that a hierarchical authority structure was needed and legitimate. Even though there were complaints about specific decisions or individuals, the organizational design was not to blame, rather, it was to be upheld. These official office practices were important mechanisms that maintained organizational design. In my study, the individuals accepted and helped to create authority differentials through legitimation of organizational design. The authority relations and work processes were legitimated and maintained through these routine perceptions and behaviors.

Unofficial Office Practices

Another part of the informal organization was the unofficial office practices that mitigated some of the negative effects of organizational design. The clerical workers developed counter-strategies to gain satisfaction, more control, and to get around the rules. The degree to which they could implement these office practices was an important factor in determining their satisfaction at work. These behaviors and perceptions were empowering. But they were also oppressive.

Goffman's notion of "secondary adjustments" is helpful in understanding these practices. In his essays on total institutions, *Asylums*, he found that inmates developed strategies, or "secondary adjustments" which were

> ...any habitual arrangement by which a member of an organization employs unauthorized means, or obtains

unauthorized ends, or both, thus getting around the organization's assumptions as to what he should do and get and hence what he should be. Secondary adjustments represent ways in which the individual stands apart from the role and the self that were taken for granted for him by the institution (1961:189).

The unofficial practices in this study did meet and further the official goals of the automated office by keeping the workers in subordinate positions. The concept of unofficial office practices was chosen because it illustrates everyday office practices that broke the rules and created a better environment for the worker but also contributed to the overall maintenance of the superordinate-subordinate system. Hence, unofficial office practices helped to create authority differentials while also being empowering. There were three general areas of unofficial office practices that related to

1) work relations,
2) work processes, and
3) company policies.

One of the unofficial office practices involved the creation of *worksite relationships*. Relationships between workers made their job more enjoyable. Many of the clerical workers said that the part of the job they liked best was working with people—their relationships with co-workers. They felt connected with their co-workers (Baker 1986). The emphasis on co-worker relations helped the operators to deal with their lack of control. This is what Kanter (1977) found in her study of employees in dead-end, low power jobs. The relationships between workers made the work itself less alienating.

Sara, a data entry operator at Sporting Goods International, stated that "What I like best about my job is the people I work with, the interaction, the feeling like I am being productive. Even though it might be menial, I am a link in a chain." Sara knows her job is menial but she makes the best of it through social relations at the office. Diana, a customer service representative at Sporting Goods International, also places emphasis on co-worker relations.

> I think my relationships with people are really good. I enjoy everybody here. There's, you know, a couple people I have personality problems with. It's just, you work around those problems. I think all of the people here have strong personalities. But we all know the personality of the other and how to work with or relate to that person.

The emphasis placed on co-worker relations functioned to make the work situation more tolerable for the clerical worker.

Another set of unofficial office practices was related to the *work process*. It was common for the clerical workers to look busy so that they would not be asked to do chores they did not want to do. This was a kind of "impression management" (Goffman 1959). They managed to be busy by spreading out their work, typing slowly, and/or getting away from the desk on company business. Similarly, I was told to slow down or there would not be enough work and I would get bored. This parallels the studies of factories where workers controlled the pace of work and "rate busters" were ostracized (Mayo 1945, Molstad 1986). Looking busy was an unofficial office practice residing in the work process.

The computer, as a product of technology, made the job more creative and the work more enjoyable. The operators could easily play with the machine without getting caught because it was an individual process.[3] The degree to which that creativity could be expressed was affected by the computer program, job description, and flow of work. Andrea, a word processor operator for Inland Attorneys, noted that "The word processor, the computer, are constantly changing and updating. And you have various different ways to get one thing accomplished. You can choose and figure out the fastest, easiest, and nicest looking way to get something done." Data entry positions were less adaptive to creativity because the computer programs were designed to just add numbers and not manipulate text as in word processing.

The third set of unofficial office practices had to do with the *breaking of company policies* in order to make the work situation more comfortable. For example, the operators would find ways to eat at their desks or stations, not write up fellow employees for being late, and exaggerate the truth regarding production levels. The receptionist at Beach Attorneys would hide her food and books under her desk and would read or eat when no one was in the room. When someone entered she would hide the food on the shelf or her book in a drawer. Or, Ernestine, at Inland Attorneys, would get "stoned" to deal with "this shit." She felt that smoking marijuana made her job more enjoyable.

The clerical workers did manage to get around some of the oppressive aspects of organizational design by creating unofficial office practices through their work relations, the work process, and breaking company policies. They did manage to mitigate some of the organizational design through the use of these empowering strategies. They developed perceptions and behaviors that made their work world more tolerable and creative. Instead of getting angry, they constructed a situation in which they could feel good about their jobs.

Alienation, Control, and Clerical Workers

The informal organization of the automated office can be seen as a reification of the organizational structure or as an alienation from society. A sociological understanding of alienation suggests that economic development and/or automation has led to greater subjective and objective alienation from work, products of work and fellow workers (Marx 1867, Erikson 1986). These components of alienation are supposedly inherent to the organizational structure in capitalist societies. This was a basic theme in the early works of Marx. Most interpretations of Marx criticize his theory for being overly deterministic and focusing too much on the material aspects of social reality. Although attempting to bring subjective analysis into the picture, Neo-Marxists still tend to approach the subjective aspects of social reality using objective sources (Marcuse 1964, Habermas 1970). Such arguments fail to take into account the creative subjective components of human nature and are limited in their explanatory utility in terms of the contemporary automated office. Although studies about automation have focused on the interrelated but theoretically separate issues of alienation and false consciousness, they have largely overlooked the question of the meaning of these processes as they are experienced by workers.

Clerical workers' perceptions and activities in my study countered the traditional arguments about automation and alienation. They did not experience a loss of total involvement in the activity of working, contact with the product of their labor, or co-workers. The following transcripts illustrate this point. These are quotes in which the workers express a sense of connection with the activity of working. In the following quote Beryl, a customer service representative at Sporting Goods International, is talking about what she likes best about her job. She likes being able to help people out. If customers call in with complaints and are frustrated, she helps them figure out the problem with their order and then fixes it.

> What I like best really, is dealing with different customers. It's, you know, it's hard to explain but it can be pretty rewarding when you do something, you help somebody out, you know. I like to go home some days and know that I just helped somebody do something. I've had people call me up and thank me for helping them do something, or whatever, and that is very satisfying.

In this quote Beryl describes satisfaction with part of her job. She likes her interaction and connection with customers. She does not sound alienated from her product—that product being service to the customers. In the next quote Sara, a data entry clerk at Sporting Goods International, talks about what she likes about her work. Sara says that she likes the challenges she faces when using new machines.

> I can run an adding machine like a breeze now, but boy, when I first started, and you know, some of these girls I was working with, I thought, God, I'll never be able to do that, you know. But you get to the point where you know, that's one of the challenges. I mean, you have to make little challenges for yourself at work, especially when we have jobs like this. Because, it's not that it

> doesn't take that much thinking as it does, well it does take thinking, too, but it takes a lot of manual skill.

When Sara first started her job she felt that she would never be able to use the machines like the women with whom she worked. But as she became more accustomed to the job, she found it challenging. Sara made challenges for herself at work, she was creative about the work process. Sara and Beryl were involved, and seemed to enjoy the activity of their work.

In the next quote Dawn, a word processing coordinator at Inland Attorneys, was justifying her position that word processing operators should not talk to each other while they are working. She was criticizing another word processing center at Inland Attorneys because the coordinator of that center let people talk. Dawn felt that talking led to more typos and documents which were not as "professional."

> I'm always strict with people who are on our floor about any document they send out. Even though it says "draft" I want it to look as if its the final draft. So no typos, and we've had very few complaints on that at all in our center where other centers have had quite a lot of complaints. So I'm, perhaps, a stickler on how a document gets sent out.

Dawn talks about the importance of the appearance of documents that leave her center. She seems to be very much connected to the end product.

Another component of alienation is said to be separation from fellow employees. The participants in my study are not "estranged from their fellow creatures" in any subjective sense. They created work relations which made dealing with their job easier. The following quote is by Kim, the word processing coordinator at Western Bank. She talks about her relationships at work. The operator is responding to a question about what she likes best about her work. Notice the level of significance placed on relationships at work.

> What I like best about my job is the variety aspect of it, that I don't have to sit down in front of the terminal typing eight hours. Also, the people that I deal with. I like to work with people...users, the operators, you know, that type thing. There's interaction with each other in the center and also with the user.

This quote is illustrative of the importance of the relationships between co-workers. The idea that alienation results in estrangement from one's fellow workers, a seemingly direct consequence of automation, is not upheld in my data. The office workers I talked to felt connected with others on the job. This connection made the job worthwhile.

The findings of my research raises some new ways to think about alienation and automation. We need to look at the consequences of automation with greater detail, more complexity, and not just as an inevitable result of corporate capitalism. The perspective of the workers' experiences, and the understanding of the work they do, help us to better understand automation and alienation. The notion that subjective alienation is a separation from the activity of working, the product of labor, fellow workers, and may be related to the introduction of new technologies into the workplace, is not evidenced in my study. The office workers talk about their work as rewarding and challenging. Worksite practices such as helping people, perfecting a manual skill, and producing a good end product are rewarding and challenging aspects of the work. My data show that office workers express a sense of connection to the work and the product even after the introduction of automation. It has also been argued that alienation is estrangement from fellow workers. This, too, has been predicted as a consequence of automation in the office. The workers I interviewed and my personal experiences illustrate that this is not the case in automated office. The workers do feel connected. They highly value the interaction with other workers and see their work as meaningful.

What I did find was that workers experienced an absence of control and a distantiation from the work process as a result of the organizational design. The problems in the office were, then, social organizational. Clerical workers experienced dissatisfaction when they could not control the work they did.

They did not control how the work got distributed, how the tasks were divided, or how they got treated by management. The following quotes illustrate the connection between organizational design and worker dissatisfaction. In the first quote Barb, a word processing operator at Beach Attorneys, discusses the process documents go through before they finally leave the computer room. She complains about the number of times she has to the revise documents, particularly large documents. Typically, a document will come into the computer center from a paralegal, then from an associate attorney, and finally from the partner. These documents can go through six drafts at each of these stages. This frustrated and created a lot of extra work for Barb. In the following quote she explains one such situation.

> Peritonitis. They spelled it wrong. I don't know, I think it was handwritten the first time and it was spelled that way in the whole document. The term appeared there and it was spelled the same way, wrong. They had me print it out again, and just the printing is time consuming. So sometimes it's frustrating. I wish they would, after making the second draft, have everybody whose on the project read it and make all the corrections before they say it's final, instead of printing a 100 page document final each time.

Becky is frustrated about her lack of control over the work process. In the next quote we can also hear frustration about lack of control over job duties. Sara is a data entry clerk at Sporting Goods International.

> Probably what I like least is when someone shoves work off on me that they could do themselves but they don't want to, well, or if someone takes me for granted. For instance, if you help somebody out and then they take it for granted that you are going to do it for them all the time. Or if they are unwilling to help you out. I would have to say that what I like the least is when someone else starts doing something a particular way and starts

little power plays. I don't like people to act like they are trying to make me feel inferior or make themselves superior to me.

Sara's talk gives us a sense that the organizational design and work process is an impediment to her satisfaction at work. She lacks any real authority to determine the activities at work which comprise a significant amount of her time. Organizational designs enhance dissatisfaction if they are rooted in a lack of worker control over the work process.

The automated office as a worksite for clerical workers is wrought with social organizational problems. The resulting ambiguity involved in what is rewarding and challenging is significant because it is an indication of the core of the problem. Workers insist that their jobs are rewarding, challenging, and creative while admitting that they lack control over the work. I am arguing that the core of the problem is the lack of control. They do not have much control over how the work gets distributed or how they get treated. They do have control over producing a good end product, helping people, and relating with co-workers. While this may not mean that their jobs are intrinsically rewarding, it should caution us against describing workers as completely alienated, degraded, and passive creatures. What seems to be happening is that they are making the best out of a bad situation through cooperative and creative worksite practices.

Through the informal organization, clerical workers construct informal associations with the products, processes, and social experiences of their work. By addressing the subjective world of the workers' ideas and perceptions I found that alienation entails being estranged from the ability and opportunity to utilize the informal organization to circumvent the constraints of the formal one. The important point of this analysis is that this estrangement is maintained and reinforced through the beliefs and behaviors of the individual workers. However, those beliefs and behaviors also relieve the workers of some of the estrangement.

Conclusion

The clerical workers' informal organization is intimately connected with the macro-social context of specialization, rationalization, and sexism. It is also connected to the authority structures and work processes contained within the organizational designs of the companies for which they worked. The macro-social context and organizational designs are two levels of social reality setting boundaries for workers' perceptions and activities. But these levels of reality were not all determining. The clerical workers' office practices, as part of their informal organization, are the level of social reality within which their perceptions and activities take place.

The official office practices are directly related to the oppressive social structures and reinforce them. These practices are based on such behavioral norms as getting the work done, liking the work, and sexism. In the unofficial practices the clerks found ways to increase satisfaction by having strong social relations, making work creative and challenging, and breaking company policies—thereby increasing their perceived job satisfaction. The link between macro-social context and everyday life is embodied in official and unofficial office practices. The individual accepts and helps to recreate oppressive structures and processes by using official office practices that are couched in a process of legitimation. But the workers actively participate in emancipating themselves from their subordinate position through unofficial office practices. Although these unofficial office practices ultimately re-create the oppressive structures and processes, they are active behaviors the workers use to make their workday better, and are not just a resignation to oppression. The comparison of office practices in these automated offices illustrates some general themes and/or ways in which specialization, rationalization, and sexism are accomplished in everyday life. Clerical workers' office practices are both constrained and empowering. Given the fluid, active, and creative aspects of humans it is not surprising that they manifest creative ways of dealing with subordination. But given the deterministic qualities of the macro-social context and organizational designs it is also not surprising that clerical workers re-create those very processes that subordinate them.

Notes

1 Wolf (1986) uses historical evidence for her argument that the link between structural power and individual consciousness occurs through a process of legitimation. Legitimation of oppression is comprised of many dimensions which are based in the perceptions of the oppressed. According to Wolf, the last and most important step in that process is "reflexive legitimation" which is comprised of four components. The first component of the reflexive legitimation process is the individual's self-conception. This self-conception grows out of social interaction with others. The second component is an objectification of one's self. The third component of reflexive legitimation is the effect of one's placement in the social universe on the formation of the self concept. The fourth and final stage is reflexive legitimation. Wolf argues that dispriviledged people know they are inferiors—reflexivity signals to them what they are expected to do, what they have become, and the propriety and rightness of their status (228-229). As a result of this process, the legitimation of oppression emerges as a normative reaction by the oppressed.

2 Coloring is a process where a specialist, through some tests, determines the best color of make-up, clothes, and the types of material that a person should wear to maximize his or her appearance. These decisions are based on the color and texture of one's skin, hair, eyes, etc.

3 Of course, this only worked in an office where there was no keystroke monitoring. Where there was keystroke monitoring, this was an empowering activity that was not available.

5

BORED AND BUSY AT WORK:

A SEMIOTIC MODEL

The interactional and behavioral patterns that comprise the automated office present a connection between the macro-social context, organizational design, and workers' activities and perceptions. The workers are bound to certain activities through organizational design. However, they are also active participants in the creation of their work world even when that work is stressful, tedious, and repetitive. Clerical workers in automated offices create and experience their work world while operating under organizational constraints such as not being able to define the work they do, how the work is done, or the time frame in which it needs to be done. The ethnographic approach of this book focuses on the everyday level of clerical work—what actually goes on in the automated office from the perspective of the participant. The clerical workers do not have control over their actions in the broadest sense, but they are, nevertheless, active participants who share in the production of the occupational world. This chapter further illustrates the notion that clerical workers attempt to and succeed at resisting some forms of domination.

Semiotic Square

A semiotic square is useful in articulating the perceptions and activities of clerical workers in automated offices. The method of a semiotic square is based on a search for opposition of meanings which occur in cultural settings. These oppositions provide the researcher with sets of possible domains of behavior and their basic features. By analyzing relationships of complementarity, contrariety, and contradiction within the setting, the researcher hopes to isolate poles of meaning (Greimas and Courtes 1982:308-311). Although semiotics is usually seen as a structural and static method, the use of the arrows in the square indicate interaction between the oppositions and help to include the notion of process and interaction. Semiotics is made a more valid approach to members' perspectives in this study by using concepts which are ethnographically valid.

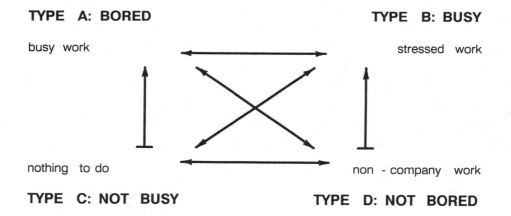

Figure 2. Bored and Busy Model

The semiotic square used here does not cover every action in the office but serves as a methodological tool to categorize and contextualize behavior. The semiotic square is useful for an analysis of the clerical workers' activities and perceptions, within broad parameters, because it is a logical presentation of semantic categories. This semiotic square is one way to illustrate the general context of office work through an examination of domains of opposition. As a result I can analyze meaning in context, and therefore, clerical workers' perceptions and activities. Using the semiotic square as a method led me to discover oppositional relationships associated with clerical workers' perceptions of work. Boredom and busyness were quite prevalent as activities and perceptions. There were times when the clerical workers were very busy and there were times when they had no work at all.

Four oppositional categories of perceptions and activities associated with boredom and busyness emerge through an analysis using the semiotic square. "Bored" (Type A) is a time during which the workers are bored with their work. "Busy" (Type B) is a time in which the workers are very busy with work. "Not Busy" (Type C) is a time when the clerical workers have nothing to do. "Not Bored" (Type D) is a time when the clerical workers do noncompany work. Types A and B are in a contrary relationship as are Types C and D. A contrary relationship is not a negation, but an opposite side of the same coin—the coin being either having company work or not having company work. Type A and C are in a complimentary relationship as are Types B and D. A complimentary relationship is one in which the behavior processes resemble each other—the processes being either bored or busy.[1]

Bored (Type A) and Busy (Type B)

Given the sociological literature on work, these behavior patterns were expected in automated offices. The clerical workers were *bored* because of the monotony of the work, and *busy* because of the pressure inherent in the organizational design. These two types of worksite practices are in a contrary relationship with each other and constitute a majority of the support staff's time. The main theme is that they are doing company work. These worksite practices differ from *not bored* and *not busy* because the "objective" situation is that they really do have company work.

Bored (Type A) Bored was the most prevalent situation for the clerical workers. It was constituted by a stretch of time in which the workers were doing monotonous or everyday kind of work. These tasks were not challenging or exciting. This was a predictable state—given the literature. What became interesting was how the workers chose to deal with boredom. Klapp (1986:11) argues that boredom is a problem people experience in different ways, "Some boredom may be in the eye of the beholder. Some is due to a social structure or culture that affects many, though participation varies." There were two types of boring experiences for the clerical workers. One was a dreary kind of boring and the other was a type of altered consciousness. Sometimes the workers experienced boredom as dull and dreary due to monotonous and repetitive work. This kind of experience was what we would expect from the literature that tells us that automation leads to rationalization and deskilling (Braverman 1974). The second type of boring experience with work was much more interesting. It was a state in which the workers managed boredom and went into a different level of consciousness.

The following excerpt from Joan's transcript, a data entry operator at Sporting Goods International, illustrates the first type of boring experience. Joan was not very happy with her position at Sporting Goods International. Her relationship with Conrad was problematic and she was concerned that she might lose her job after the transition to the larger computer.

P: what is it really like for people who work in offices?
J: boring
P: boring?
J: boring, that's it
P: that's how you feel about it huh?
J: yea, it's boring
P: what about different kinds of jobs that you do?
J: well I guess I'm bored, I'm just bored right now
P: is it bored with what you actually do, or the people or??
J: with what I do, it's the speed you have to do everything real quick and then

you don't have anything to do and I am sitting there trying to figure out what to do, I just get bored

This dialogue illustrates the first type of boredom.which centers around its mundaneness. We get a sense of resignation. This differs from the next two quotes because it does not articulate an active response to the boredom. The following quotes are examples of computer operators talking about the second type of boredom—an altered state of consciousness.

In the next quote Randy, a word processing operator at Beach Attorneys, describes the mechanical aspect of the job.[2] Just before this segment of the transcript he was talking about the hardest aspect of the job which is doing what someone else tells him to do. He says that he has to control himself from getting angry when told what to do in order to get through the assignment. But the easiest part of the job for Randy was the mechanical nature of the task. He describes the second type of boredom.

> The easiest, the very easiest aspect of the job as a typist or someone who transcribes dictation tapes or whatever you want to call it is its completely mechanical function. It requires no brain activity other than some nerves that have been strung up now between my fingers and my ears. As a matter of fact another operator was kidding me the other day. I had closed my eyes while I was typing and listening to dictation tape and I have almost gone to sleep doing this and at the same time typing and other operator started laughing at the fact that I was typing so fast with my eyes closed completely ignoring the screen in front of me. I have almost dozed off in that condition and I think it's funny. It shows you how, you know, a skill that goes beyond intelligence becomes just plainly mechanical.

In the next quote Jennifer talks about a similar kind of experience. She called the second type of boredom "veg out"—which refers to the nonthinking aspect of vegetables. Jennifer is responding to a question about what she likes best and least about her work.

I mean this keying in and billing people is not fun. It's kind of a thing, it's kind of nice because you don't have to think all the time if you just want to veg out you get to the point where you can do it and talk. In there, in the computer room, we are talking all the time because it's such a tedious chore.

Both Randy and Jennifer create an almost enjoyable experience. They find ways to get around organizational constraints.
 Csikszentmihalyi (1975) talks about a "flow experience" at work or at play in which the person experiences a total involvement and merging of action with awareness. People who perform this can temporarily forget their identity and problems. The state of flow does not depend on the objective situation but on one's perspective of challenges and skills. The phenomena which Csikszentmihalyi refers to is similar to the descriptions given by Randy and Jennifer. Even though the work is dull and boring, they can improve it.
 Molstad in an article "Choosing and Coping with Boring Work" demonstrates how workers in an industrial brewery cope with boredom by fantasizing and daydreaming.[3] Through four years of participant observation Molstad found that the workers attempted to avoid boredom by mentally escaping the work. He mentally escaped the work through fantasies about where the bottles of beer were going and who was going to be drinking them. Molstad recalls his experience as a kind of "mental regression" in which his altered consciousness became more childlike and harder to shake off as time went on. In my study Randy and Jennifer obtained a consciousness similar to what Csikszentmihalyi calls a "flow experience" and Molstad refers to as day dreaming and fantasizing. In response to being bored with work they "veg out" or almost go to sleep. These were active responses to the mundaneness of computer entry.
 The experience of being bored with company work was managed in various ways. The workers responded to the inherently repetitive and tedious everyday work by an inactive boredom as in Joan's case. Or, the workers re-

sponded by creating a state of "flow" that took them out of the task and into talking or sleeping. These were two types of responses to work.

Klapp says that boredom is socially constructed. He argues "...that once named, "boredom is not just a state of feeling but a concept, a role, a socially constructed reality. The very name might suggest being bored; and the role might carry the obligation that one ought to be bored in certain company or circumstances. Hence, boredom could become a fashion, or a pose such as that of the blase aristocrat or of the romantic bohemian" (1986:33). The second way to deal with boredom was to make it a "fashion" and, as such, the clerical workers made the situation better for themselves.

Busy (Type B) In the busy (Type B) situation, the workers were trying to get the work done—fulfilling the organizational demands. Following the semiotic square, Type B "busy" was in a contrary relationship with Type A, "bored." This does not mean it is a negation but rather a relation that seems opposite. Busy and bored differ because busy was a time when there was an undue amount of stress, and bored was a response to repetitive and tedious work. Both of these sets of circumstances were times when the clerks did company work, but they took on qualitatively different characteristics.

Busy was closely related to the formal organizational design. One of the strongest strategies the bosses used to maintain control and subordination of the clerical workers was the manipulation of the work process and the work flow. It was not unusual for the bosses to disregard whatever the employees were doing and demand that other work be done, often within an unrealistic time frame. The bosses did not give much concern to the workers' official job descriptions and personal interests. The design of the work process did not take the needs of the operators into consideration. Theoretically and in reality there were times in which there was no work, and there were times in which there was too much work. The operators had little control over the amount of work because the users defined the flow of the work and management decided the structure of the work process. The bosses typically insisted that the work had to be done yesterday, when they handed in documents to be processed. I would speculate that the clerical workers would have preferred to have the work spread out more equally. Instead of having really busy days like during the end of the month rush or really slow days, there would be a more even work flow.

One way the clerks were busy with work was when they were required to stay late, come in early or on the weekends. At Beach Attorneys there was one incident where the support staff was called in at 3 a.m. to xerox a client's files because they were being petitioned to court the next day. Although this is an unusual example, it highlights the point that clerical workers can not define the use of their own time or the structure of the work process. The clerks would try to prevent such stressful times by keeping caught up with the routine work. However, since they had little control over how much or when the work came into the center they would, systematically, be put in situations of extreme anxiety.

Another example of busy with work in an automated office was the end-of-the-month rush. This took place in most offices because reports and bills were generated once a month. It did not really matter if the everyday work was caught up, the end of the month was always stressful. In the next quote Jennifer, the lead computer operator at Sporting Goods International, responds to a question about what she likes least about her job, "trying to get a lot of orders out and then after you get all these orders out and you work all this overtime you have to turn around and do the month end closing and you have tons of billing. Like you'll have just stacks and stacks of stuff for billing and the management wants all their sales analysis reports and all that kind of stuff." Jennifer is very busy at the end of the month, due to the organization of the work flow and her lack of control over it.

The final way in which clerical workers experienced being busy with work was a result of the bosses' hostile or erratic behavior that could cause tremendous amounts of stress. During my employment as a data entry clerk I was required to cover for the receptionist while she was at lunch. I found this position one of the most stressful in the whole project. Throughout my fieldnotes I describe periods of extreme stress relating to how to use the phone equipment and how to get the calls back to the attorneys while taking care of the people in the lobby. There were many occasions when I was criticized for not taking a message or not responding to a phone call in the correct manner. I would actually be scared if certain attorneys would call me for some kind of information. Hostility and degradation made busy with work more stressful.

It was hard for the employees not to be anxious and pressured as a result of the structure of the labor process. There were many times when the clerical workers were very pressured and busy with work. This was a result of formal organizational designs which gave the clerical workers little control over the labor process.

Bored and busy were contrary states that constituted the majority of the clerical workers' time. Bored was managed in two ways—being dull and dreary, and as a "flow state." Busy was less easy to manage because it was a time of anxiety and pressure resulting from working overtime, end of the month rushes, and bosses' hostile behavior. Klapp (1986:2) says that a paradigm of information society shows that "meaning and interest are found mostly in the mid-range between extremes of redundancy and variety-these extremes being called, respectively banality and noise." For the automated offices in this study, Klapp's analysis translates into being bored with routine work, or busy with excessive and unpredictable work.

Not Busy (Type C) and Not Bored (Type D)

Not busy (Type C) and not bored (Type D) were both part of the clerical workers' informal organization I did not expect because it was unexplored by most of the literature and not talked about by the clerical workers. I expected clerical workers to have quite a bit of work to do, whether boring or not. It was not until my phase of participant observation that this category of behavior became apparent. Not having enough work to fill their time turned out to be an occupational secret known primarily to the clerical workers. This was part of the "hidden" informal organization and kept as a secret due to fears of an increase in work load or job loss. Therefore, if the clerical workers talked about what really happened—having no work—they would be acting against their own and other clerks' unwritten rules. Beckford (1978:251) says that

> Speakers' decisions about the appropriateness of different resources are made partly by reference to their sense of the kind of context in which they are speaking. And one of the considerations to be taken into account in making such decisions is the set of rules which are considered to govern speech and action in particular contexts. It is also assumed that. speakers are concerned to display their knowledge of the rules for the sake of proving their competent membership of the relevant social groups.

Beckford's notion that what people say is a reflection of the social context in which they are speaking, illustrates the clerical workers' norm to not talk about having no work. As a result, being not busy and not bored were complicated managed accomplishments in the occupational world of small to mid-size offices. Not bored and not busy are more difficult accomplishments in the automated office because the clerical workers are operating under organizational designs and macro-social constraints to keep production and profit up.

Not Busy (Type C) Not busy (Type C) constituted the least prevalent situation for clerical workers and was essentially a transitional phase during a time when there really was nothing for the clerks to do. This situation could arise at any time and was not necessarily linked to the workload. Usually this would happen just before the employees left the office or in short periods before they would find company or noncompany work. Cultural norms permitted employees not to work at particular times, including Friday afternoons, right before lunch, and at quitting time. Although it was usually satisfactory not to be working fifteen minutes before closing or lunch, it was imperative to be at or around the station or desk in case any urgent work came in. If they had no work at other times, they would have to quickly find some work. Again, this is due to fear of being assigned some unpleasant work or job loss.

Managing having nothing to do was difficult. It was mandatory for the receptionist at Coast Attorneys to look busy at all times. If the phone did not

Bored and Busy

ring she, in actuality, had nothing to do. By company policy, looking busy meant that she was not allowed to do activities that were not work related. She could not read, eat, or take personal phone calls while at the front desk. Never-the-less managing having nothing to do was an important lesson to learn because when clerks had nothing to do and management noticed it, tasks would be assigned that the clerks did not want to do. When I was working as a data entry operator, managing having nothing to do was hard to master. But, it was an important lesson because when I had nothing to do and the bosses noticed it, I would be assigned tasks that were usually unpleasant. For example, I would have to pay my supervisor's car payment, go to pick up supplies, go to the basement to organize files, etc.

Within the informal organization of clerical workers, it was normative to not be working fifteen minutes before closing or lunch. Not busy (Type C), at other than these normative times, got managed in some fascinating ways and turned into not bored (Type D).

Not Bored (Type D) Not bored (Type D) was a state of impression management—to look busy when there was no company work. It was characterized by a time when the employees were doing work that was not company related. The workers did not have any or did not want to do any company business so they did other activities such as walking around, talking, reading, and eating. These worksite practices grew out of being not busy, and were states of impression management—to look busy when there was no company work. The accomplishment of not bored constituted a very innovative use of time. The individual's management of having no work was complicated because of the organizational constraints to keep working. They did not want to get caught having no work because that would mean they would be given more work. As a result, the accomplishment of not bored was a very sophisticated use of time. The management of having no work was complicated because of the formal organizational constraints—to keep working. They did not want to get caught having no work because that would mean they would be given more work. So there was a secrecy ritual that occurred without the bosses' knowledge. Managing to appear busy while having no work was a very important lesson and one that was learned early.

One way to deal with having nothing to do was to leave one's desk, walk around, and get coffee. Under these circumstances it was hard for the bosses to determine if the workers were doing nothing or doing something. One of the most widely used strategies was to walk around and talk to co-workers. In most cases the employees would act like they were doing something other than just walking around. If a supervisor or boss would come by, they could stop in mid-conversation, change topic to a work related one, and continue talking as if they were discussing work related issues all along.

The receptionist at Beach Attorneys was a master at being busy with no work. There were long periods of time when she would have nothing to do. If the phone did not ring and there were no clients in the lobby she did not have anything to do. She was directed by her supervisor to look busy because she was always on stage. Looking busy for the supervisor meant that the receptionist was not allowed to do activities which were not work related. She could not read, eat, or make personal phone calls while at the front desk. She did, however, learn to do some things without being caught. She would put the sandwich under the shelf and her book in a drawer. She would pull them out when there were no clients in the lobby and the attorneys were in their offices. She was passing as busy. It is interesting to note that what the bosses meant by looking busy for the receptionist was meant doing nothing at all. The receptionist, however, had strategies to make the situation more tolerable.

Managing to appear busy while having no company work was an innovative procedure for the clerical workers. This situation was a cover up for having or doing no company work. It was a ritual of secrecy because the workers knew if they were caught they would be given more work.

Not busy and not bored were contrary circumstances that constituted a significant minority of clerical workers' time. The set of behaviors were not apparent at first and were not expected given the literature. Not busy (Type C) was a time when the workers were actually doing nothing. These times usually occurred before leaving the office or in the interim to finding or being given work. Not bored (Type D) was a time when the clerks were managing having nothing to do and would do noncompany activities. An understanding of ways to manage being not bored and not busy was integral to the management of clerical workers' behavior.

Clerical Work Style

Clerical workers exist in an occupational setting in which they must appear to be continually busy while at times there are no tasks to perform. The management of boredom while maintaining an appearance of busyness is a very important part of the clerical workers' daily life in the computerized office. The performance elements contained within this impression management are a clerical work style which is deemed worthy by both the clerks and management. However, a sociological perspective necessitates an analysis of the larger social, political, and economic context. It is important to go beyond the semiotic system as it applies to the particular domain discussed above and examine the position of clerical work within the larger cultural context. A clerical work style is grounded in the interaction between formal organizational control and the employees' production of their occupational world. The impression management of boredom and busyness is sociologically important because its creation and uses are a manifestation of a dialectic relationship between organizational design and workers' struggle for control of their world.

The impression management of boredom and busyness in the offices included in this study is related to organizational design and an unpredictable flow of work. The companies in this study were small to mid size and unpredictability was a result of such variables as nonstandardized customer demands and unforeseeable management needs, taking place with a relatively small staff. For example, at Sporting Goods International there was an erratic amount of customer orders and bureaucratic requests coming from overseas management. The word processing centers at both Inland Attorneys and Beach Attorneys had irregular flow of work stemming from the attorneys' client load and their cases' positions in the legal process. Such unpredictability becomes a problem for the clerical workers because the company's main priority is to keep profits up and costs down. Within this system, overhead costs including support staff salaries, must be kept as low as possible. As such, there is a push by top management to reduce personnel if the supervisors can not schedule the work flow so the workers are always busy. The kind of organizational design creates an occupational setting in which the clerical work style is characterized by the management of boredom while maintaining an appearance of busyness. The impression management discussed above was structured in accordance with organizational pressure to keep working, due to threat of an increase in job duties or job loss.

Organizational designs which gives control of the authority structures and work process to management creates a situation in which the clerical workers are neglected. It was not unusual for management to disregard whatever the employees were doing and demand that other work be done, usually within an unrealistic time frame. The workers' official job descriptions were disregarded by management. The work process was not developed by the clerical workers and or designed with their best interests in mind. There were times when there was too much work and others when there was too little work. The operators had little authority over the work process. Unscheduled time for the support staff, unlike management, meant the elimination of positions. The clerical workers' struggle for control through an appearance of busyness was protective. The only acceptable performance was always being busy. Therefore, if the clerical workers talked about what really happened—having no work—they would be acting against their own and other clerks' written rules. The occupational world which was both boring and busy was a manifestation of the reciprocal nature of organizational design and informal organization. The management of boredom and busyness was an important thematic focus because it represents a connection between the macro level processes of control and subordination and everyday life in an automated office.

Conclusion

Clerical workers in automated offices operate within an informal organization which can be categorized as being both boring and busy. Boredom and busyness are important because their creation and uses illustrate a dialectic relationship between structural constraints on workers and their struggle for control. The semiotic square was useful for describing this situation. By looking at the cultural oppositions of being bored and busy at work we can typologize domains of behavior and their basic features. The use of ethnographically valid concepts allowed for a logical presentation of meaning within the general context of office work. This analytic tool was useful because it enabled me to locate activities which mask the fact that, at certain times, clerical workers had too much to do and at other times they had nothing to do. The clerical work "style" was dominated by the management of boredom while maintaining an appearance of busyness. It is important to look at how actors create and experience their world because until we do more than statistical, historical, and theoretical analyses, we are only making assumptions. By looking at the cultural oppositions of being bored and busy at work I can typologize oppositional categories. This was useful because I was able to locate activities which masked the fact that, at certain times, clerical workers had too much to do and at other times they had nothing to do.

The clerical workers manage boredom through a lack of concentration on dreary tasks. They can enter into a "flow" state or a "fantasy" to open with boredom (Csikszentmihalyi 1975; Molstad 1986). Or they manage being bored with no work by finding ways to look busy. In this small way, clerical workers were able to resist domination by mitigating the effects their lack of control over the labor process. Boredom and busyness are important because their creation and uses illustrate a dialectic between structural constraints on workers and their struggle for control. Just as the clerical workers used unofficial office practices to make their work situation more tolerable, they also created ways to look busy so they would not be given any more work or lose any positions. The bosses tried to gain control, and the workers tried to get autonomy.

Employees were doing activities which were not company related. This was a time when the workers did not have any company business so they did other activities such as walking around, talking, reading and eating. This worksite practice grew out of being bored with no work. Managing to appear busy while having no company work was a complicated procedure which helped to ensure the maintenance of personnel and job descriptions.

Clerical workers in automated offices operate within an informal organization which can be categorized as being both boring and busy. This situation constitutes a dialectic between historical and social constraints, organizational design and the informal organization of automated offices. Boredom and busyness are both objective and subjective phenomena. Objectively, the authority relations and work process design create a situation of inherent boredom and busyness due to tedious and repetitive work. But that description is too simplistic to describe boredom and busyness in the automated office—workers are active creators of their work environment. Subjectively, clerical workers accomplish circumstances of being busy and being bored to their own ends—to make the situation more comfortable.

cial institutions. The initial questions raised about clerical workers' knowledge system were:

1) Why don't the workers revolt?
2) How are the macro-structures reproduced?
3) Why do participant's reproduce their own subordination?
4) How is inequality within the class structure maintained?

While these questions are not new, the nature and source of the answers contained in this book are. Clerical workers' perceptions and activities in automated offices were the source of data. An analysis of cultural forms shows the participant's role in creating and re-creating specialization, rationalization, and sexism The analysis also shows that the clerical workers' informal organization is used to diminish the effects of these macro-social structures and processes. The problem with most theories and research methods which address subordination is that they exclude any discussion of knowledge systems. A sociology of knowledge theoretical framework and an ethnographic method were the most effective for analyzing control and subordination in the automated office.

Findings and Contributions

Clerical work exists within a social reality comprised of a macro-social context, organizational design, and an informal organization. The macro-social context is comprised of a market society, rationalization, and patriarchy. These macro-structures all share an exploitative nature while setting the stage for specialization, rationalization, and sexism. Organizational design of the office implements specialization, rationalization, and sexism through authority structures and work processes. However, the informal organization and work-site practices are where specialization, rationalization, and sexism are actually accomplished. An analysis of the informal organization offers us rich and new information about the role of clerical workers in the macro-social context, organizational design, and informal organization.

The macro-social context, organizational design, and informal organization of automated offices intersect in some some very interesting ways. Basically, the bosses try to gain control and the workers try to get autonomy. The bosses use authority structures and work processes as means for gaining control. A strict authority structure was particularly problematic for the clerical worker. In cases where the workers were given autonomy, they experienced more job satisfaction. While organizational designs are strategies to implement control over the worker, it is within the informal organization that workers try to get autonomy. Without a focus on the informal organization, we would not be able to see the active and creative aspects of the workers.

The clerical workers used the informal organization as a way to get autonomy. There were two models of worksite practices which exemplified the clerical workers' informal organization:

1) Official and Unofficial Office Practices, and
2) Bored and Busy Semiotic Square.

The official office practices were those in which the clerical workers directly reinforced and re-created the power relations. The unofficial office practices were clerical workers' counter-strategies which mitigated the effect of the macro-social context and organizational design. Although the unofficial production practices lessened the perceived impact of lack of control, the objective authority relations remained the same. Clerical workers got around some oppressive aspects of work but also re-created their own domination.

The second model of worksite practices was the analysis of bored and busy using the semiotic square. The clerical workers used these practices to gain autonomy and control. Clerical workers are subject to a work environment in which they have little control over the pace or content of their work. They guided their actions according to organizational design, however, they were not robots. The impression management to look busy was a means to make their situation more comfortable. In this way they side stepped the threat of more work duties and loss of personnel.

Conclusion

Analyzing the informal organization also showed that sociology can not assume such processes as alienation a priori. While it has been argued that alienation is the loss of contact with the product of one's own labor, fellow employees, and self, my data show that office workers express a sense of connection even after the introduction of automation. The problems in the office were social organizational. Workers experienced an absence of control and a distantiation from the work process as a result of organizational design. The workers admitted they lacked control over their work. They did not have much control over how the work was distributed or how they were treated. However, they did have control over producing a good end product, helping people, and relating to co-workers. They were able to find good things about their jobs, while being in an organizational arrangement that was not good in any objective sense. While this does not mean that their jobs are intrinsically rewarding, it should caution us against describing workers as alienated, degraded, and passive. What seems to be happening is that they are making the best out of a bad situation through cooperative and creative worksite practices.

In sum, inequality in the class structure is maintained through the reciprocal and constitutive interaction between the macro-social context, organizational design, and informal organization. The reciprocal nature of empowerment and organizational control in the automated office begins to answer the questions raised early in this book. The macro-social context is reproduced through organizational design and informal organization. The workers' knowledge systems were grounded in macro-social context and individual action. The participants reproduce their own subordination through processes of reflexive legitimation (Wolf 1986) and secondary adjustments (Goffman 1961). Even though clerical workers find ways to diminish the effects of their lack of control, they reproduce the very structures and processes which hold them down.

Larger Social and Sociological Issues

The good news is that some of the clerical workers' worksite practices create little victories which make them able to more comfortably manage their situation. The bad news is that there are big defeats because worksite practices operate as avenues which keep the power structure in tact.

The traditional Marxist framework does not explain control and subordination on an everyday level in the automated office and my data creates an uncomfortable graft with other theories. Even though I did not find robot-like workers, there was still something terribly wrong in the automated office. Various forms of subordination and inequality, ranging from low pay and no promotional ladder to verbal and sexual harassment, exist for clerical workers. Ironically, the very behavior which makes their jobs more comfortable also serves to keep them in subordinate positions, and prevents them from collective action.

Collective action would be the best strategy for improving the position of clerical workers. Why is collective action not happening when the need for it appears huge? As I see it, there are three barriers to collective action. The first, and most obvious, barrier is the corporations. There is little or no incentive in the economic structure for them to change. The second barrier is the legitimation process which gives the workers little incentive to change. This process pacifies the workers and is kept in place by social interaction. The third barrier to collective action is the workers' view that personalities are the root of problems in the workplace. They see the overall structure of the organization as valid. One could hardly expect them to organize around what they see as individual problems. There does not seem to be much incentive for either the organizations or the workers to improve the situation. As such, the obstacles to collective action are numerous and intertwined. In order to improve women's position in clerical work there has to be change on all three levels of social reality delineated in this book. Transformation of clerical work is not likely unless the macro-social context, organizational designs, and informal organization change. The processes and structures described in this book are oppressive and the only way to change them is through social interaction. To answer Marx, until the caged bird stops singing, there will be no revolution.

Future Research

My work just begins to scratch the surface in an analysis of the informal organizational world of clerical workers in automated offices. One direction for future research is to take clerical workers' knowledge systems more seriously for policy changes and implementation. Further ethnographic analyses would contribute much needed data to any current sociological and public policy debates including those on deskilling and comparable worth. We need to develop theories and methods which do not flatten individuation. We also need theories which address how organizational control prevents empowerment and subsequent positive experiences for employees.

Another direction for future research is an explication of different kinds of power and authority relations. A useful and insightful way to analyze power and authority is through an analysis of the accomplishment of them. A comparative analysis of different kinds of organizations would be interesting and allow us to see other instances of power and authority on the socio-cultural level. Comparative ethnographic studies would allow us to do theory building about power and authority.

A Final Note

Clerical workers in today's automated office are victims of subordination on various levels of social reality. Subordination in the office is the embodiment of the dialectical and reciprocal relationship between macro- and micro-social structures and processes. Their social and cultural worlds are both subordinate and empowering. Given the fluid, active, and creative aspects of humans it is not surprising that clerical workers manifest creative ways of dealing with subordination. But given the quasi-deterministic qualities of the macro-social context, it is also not surprising that subordination is internalized, and lived out in everyday life.

Degan (1987) says that we need to create "pragmatic assumptions" in order to lead toward positive social change for women. My research is based on the notion that the clerical workers' empowering worksite practices make no objective change in power relations. Things could be made better for the clerical worker if they were given more autonomy. Workers are not robots to the social structure, but are actively working towards creating a more comfortable workplace. My pragmatic assumption is that until the structure of organizations and knowledge systems change, clerical workers and women in general, can only have strategies which mitigate and reinforce power differentials, not change them.

Notes

[1] This theoretical framework and, specifically, the notion of cultural forms as both a source and activity of knowing, was worked out in graduate seminars in the sociology of knowledge at the University of California, San Diego. Those seminars were led by Bennetta Jules-Rosette during 1986-1989.

APPENDIX A

METHDOLOGICAL NOTES

This book is a comparative ethnographic analysis of four automated offices. The findings are based on a study conducted in San Diego, California between June 1984 and December 1985. The methodology employed was a combination of qualitative methods using interview, participant observation, and observational data collection strategies. Most of the details of the research design are woven throughout the main text of the book. What follows is a description of the overall research design.

Observational studies were conducted at two San Diego word processing centers, a bank (Western Bank) and a large law firm (Inland Attorneys), and participant observation was done as a data entry clerk in a law firm (Beach Attorneys). Interviews were conducted with persons involved in a multi-national corporation (Sporting Goods International). During the observation and interview stages of the research the focus was on management/worker relations, social organization, and the cultural world of clerical workers.

This methodology was choosen because it permits gathering data on clerical worker's everyday life. It also permitted me to experience that everyday life. I chose the everyday life perspective because it put action into the subjects that statistical and historical analyses take out. The aim of this study was an understanding of clerical work in automated offices as lived experience.

The focus of this research project is on the members' perceptions and activities. This focus necessitates a data-driven, inductive, and counter-inductive data collection method. A "bottom-up" model such as this one can add insight into areas of social life which a deductive, hypothesis testing approach can not. It is difficult for studies which use random and representative samples to address the participants' perspectives, activities, interactions, and worksite practices. Therefore, they were not appropriate.

I began this research project interested in human-machine interaction and how computers had changed clerical work. As I collected data, I realized that there was not a close fit between the sociological literature on clerical work and my data. I had assumed that I was going to find women who hated their jobs and that the work they did would be low skilled and unenjoyable. However, that is not what I found. I almost immediately came to understand that these women like their jobs and that the tasks they performed were not easy. My early focus gave way to an interest in the creation and re-creation of relations of subordination.

Because this project was data driven, I had to become familiar with computer jargon and computer skills. As a graduate student there were periods in which I entered the temporary clerical work force. It was during these periods that I became familiar with office procedures. Moreover, I had become familiar with computer skills, primarily word processing, on various computer systems at the University of California, San Diego campus. It was during this phase of the research project that categories for investigation were generated and substantiated.

The analysis contained in this book constitutes the results of an investigation into word and data processing centers which staffed and managed primarily by women. The four firms in this study which were selected using an "opportunistic" approach (Riemer 1977) and a snowball sample. The results of this investigation are not meant to be representative. The data for this research project came from four firms from which I could gain entry for interviews, observation, or employment.

There were three stages of data collection. In the first stage I studied a law firm and a bank using similar techniques. I would first submit a research proposal to a manager who would then get final approval from the Board of Directors. Then I would start observations of the word processing center.

Appendix A

During the periods of observations I would set up interviews with the operators and perform those interviews at lunch time or after hours. I did ten interviews and 20 hours of observation of these two settings.

The second stage of data collection was at a law firm as a data entry operator. I worked there for six months and conducted two interviews. It was during this stage of data collection that I was able to collect data regarding office practices.

The final stage was a series of ten interviews with persons involved in an automated office at the main administrative branch of a sporting goods corporation. The office was involved in the conversion to a mini computer. The occupations of the people varied and included secretaries, data entry clerks, and customer service representatives.

Primary topics on the interview guide were the feelings and attitudes of the operators about the computers, themselves, their bosses and their jobs (Appendix B). It was during the first stage of observation and interviewing that I discovered the operators liked and were proud of their work. I was also learning about the organizational designs of the offices—the authority relations and work process. During the second stage of participant observation I was particularly interested in the subjective qualities of work in an automated office. It was during this phase that dialectical nature of subordination and the empowering strategies of the operators became more apparent. The final round of interviews substantiated my understanding of the data about organizational design and informal organization that I learned in the first and second stages, respectively.

In designing my research strategy, my concern was to understand the perspectives of the clerical workers and the managment for whom they worked. The recognition of multiple realities (cf. Schutz 1967) was the guiding principle in my participant observation and interviewing techniques. The notion of multiple realities and knowledge systems was the basis for my choice of analytical methods. My data were subjected to sociolinguistic methods in order to study how things were said. The data were also subjected to semantic analysis to understand and categorize the meaning of what was said.

Taken together, these strategies allowed me to analyze mechanisms producing and reproducing control and subordination.

During all phases of data collection, I would continually write field notes and expand on them when I left the office. I would also talk into a tape recorder after leaving the setting to note my feelings and observations of that particular day and how I felt about the research project as a whole. I utilized these techniques of data collection to ensure that I had the best possible recall.

I would tape record the inteviews and then later transcribe the tapes for analysis. I performed various types of data analysis on the interviews and field notes. One type was content analysis. I looked at what were the defining features of the office for the clerks. Another was semiotic. I looked at cultural oppositions. Another type of analysis was the explication of the subjective understanding I obtained from my own participant observation.

Data analysis was an ongoing process from the start of data collection to the finish of the write-up. My data took the form of transcribed interviews, fieldnotes, and observations. Refer to chapter four for many of the details regarding entry, length of study, and number of interviews within each setting. The data analysis started with the use of discourse and conversational data. The sociolinguistic analysis allowed me to find forms of talk which categorized the speakers as well as the information discussed. For instance, upper level management often refer to the word processing staff as "word processors" rather than "computer operators." This mixing of linguistic categories between human and machine is a great source of complaint and operator dissatisfaction. Consequently, I was able to measure dissatisfaction from both the content of what was said and the manner in which it was discussed.

This combination of methods allowed me access to the perceptions and activities of the workers as those perceptions and activities relate to macro-social context and organizational design. It was necessary for me to experience as well as interview and observe to get the most valid data to answer my research questions about clerical workers in automated offices. I was able to gain a better understanding than would have been possible with just one data collection strategy. As a result, the project was informed by ethnographic data and a subjective understanding of clerical work.

APPENDIX B

INTERVIEW GUIDE

I. Description of research project

This project is designed to collect data from those persons involved in recent office automation.

I am primarily interested in **your story** about your work and office automation.

   a. the goal of this project is to define what is important to those involved in office automation.
   b. consent form
   c. any questions
   d. format of interview, four sections
      1. life historical
      2. work world
      3. skills and tasks
      4. office automation

II. Life Historical

What is your name?

How old are you?

What was important when you were growing up?

What was your schooling like?

Tell me about yourself today?

What family situation are you in today?

III. Work World

How do things get done around here?

What goes on during an ordinary day at work?

Would you explain your occupation to me?

What do you like best about your job?

What do you like least about your job?

What is the hardest thing about your job?

What is the easiest thing about your job?

What kinds of relationships do you have with the people you work

What kinds of relationships do you have with the people you work for ?

What would a day in the life of your work utopia be like?

IV. Skills and Tasks

What are the tasks that you perform?

What skills are required to do those tasks?

How did you learn to do those skills?

Where did you learn to do those skills?

How long did it take you to learn them?

Appendix B

V. Office Automation

How do you feel about office automation?

What effect do you think automation has on your work related activities?

What is the hardest thing about office automation at work?

What is the easiest thing about office automation at work?

How do things get done now compared to before?

What will the next stage be?

What happened when people learned about acquiring the new machines?

How are you learning to use the machine?

Corporate Culture and the Automated Office

University of California

Consent to Act as Human Subject

Principle Investigator:
Phyllis L. Baker

Name _____

Date _____

This form is designed to explain the strict confidential nature of this project to its participants. The information collected during this interview will be available only to the principle investigator. Any of the data or subsequent findings of this project will be used only for social science research.

This project will employ the following techniques to secure the anonymity of any possible marks of identity relating to either you, the participant, or your workplace. Any marks of identity will be changed or deleted in both the transcript and during any presentation of the data. This will include any names or locations of people or places that you may mention during course of this interview.

Please read and understand the following guidelines for this interview:

1. You do not have to answer any questions at any time or for any reason if you choose not to.
2. I will answer any questions you may have regarding this project, the interview process or anything that maybe of concern or interest to you.
3. You may terminate this interview at any time.
4. Finally, this interview will be audio-taped and hand notes will be taken.

I hereby authorize _____ to interview me about my life and work experiences for **Corporate Culture and the Automated Office** project. Furthermore, I understand the above stated guidelines of the interview process and the techniques which will be employed to ensure my confidentiality.

REFERENCES

Attewell, P. (1987). The De-Skilling Controversy. *Work and Occupations*, 14(3), 323-346.

Baker, P. (1986, June). *The Automated Office: The Feminization of a New Technology*. Paper presented at National Women's Studies Association.

Bandura, A., Ross, D., & Ross, S. (1961). Transmission of Aggression Through Imitation of Aggressive Models. *Journal of Abnormal and Social Psychology*, 63 3 , 575-582.

Barker, J., & Downing, H. (1985). Word Processing and the Transformation of Patriarchal Relations of Control in the Office. In D. Mackenzie and J. Wajeman (Eds.), *The Social Shaping of Technology* (pp. 147-164). Philadelphia, PA: Open University Press.

Barnard, C. (1938). *The Functions of an Executive*. Cambridge, MA: Harvard University Press.

Beckford, J. (1978). Accounting for Conversion. *British Journal of Sociology*, 29(2), 249-261.

Berger, P., & Luckmann, T. (1967). *The Social Construction of Reality*. Garden City, NY: Anchor Books.

Blau, P. & Scott, W.R. (1962). *Formal Organization : A Comparative Approach*. San Francisco, CA: Chandler Publishing Company.

Blauner, R. (1960). Work Satisfaction and Industrial Trends in Modern Society. In Walter Galenson and Seymour Martin Lipset (Eds.), *Labor and Trade Unionism* (pp. 62-68). New York, NY: John Wiley.

Blumer, H. (1962). Society as Symbolic Interaction. in Arnold Rose (Ed.), *Human Behavior and Social Processes: An Interactionist Approach* (pp. 78-89). Boston, MA: Houghton Mifflin Company.

Braverman, H. (1974). *Labor and Monopoly Capital: The Degradation of Work in the Twentieth Century.* New York, NY: Monthly Review Press.

Burawoy, M. (1979). *Manufacturing Consent: Changes in the Labor Process Under Monopoly Capitalism.* Chicago, IL: University of Chicago Press.

Cicourel, A. (1964). *Method and Measurement in Sociology.* New York, NY: The Free Press.

Clifford, J. (1988). *The Predicament of Culture: Twentieth-Century Ethnography, Literature, and Art.* Cambridge, MA: Harvard University Press.

Clifford, J., & Marcus, G. (Eds.). (1986). *Writing Culture: The Poetics and Politics of Ethnography.* Berkeley, CA: U.C. Press.

Collins, R. (1986). *Weberian Sociological Theory.* Cambridge, MA: Cambridge University Press.

Combs, S., Phaneuf, S., & Flynn, P. (1985, December). *The Sociology of Knowledge in a Penguin Ethology Laboratory.* Paper presented at Colloquium on Ethnographic Approaches on Reasoning, Production of Knowledge and the Process of Discovery in Scientific Laboratories, Paris, France and L'Ecole Polytechnique.

Cromptom, R., & Jones, G. (1984). *White Collar Proletariat: Deskilling and Gender in Clerical Work.* Philadelphia, PA: Temple University Press.

Csikszentmihalyi, M. (1975). *Beyond Boredom and Anxiety.* San Francisco, CA: Jossey-Bass Incorporated Publishers.

Davies, M.W. (1982). *Woman's Place is at the Typewriter: Office Work and Office Workers 1870-1930.* Philadelphia, PA: Temple University Press.

References

Deegan, M. (1987). Working Hypothesis for Women and Social Change. In Mary Jo Deegan and Michelle Hill (Eds.), *Women and Symbolic Interaction* (pp. 243-249). Boston, MA: Allen and Unwin Inc.

Downing, H. (1980). Word Processors and the Oppression of Women. In Tom Forester (Ed.), *The Microelectronics Revolution: The Complete Guide to the New Technology and its Impact on Society* (pp. 275-287). Cambridge, MA: MIT Press.

Erikson, K. (1986). On Work and Alienation. ASA, 1985 Presidential Address. *American Sociological Review*, 51, 1-8.

Ferguson, K. (1984). *The Feminist Case Against Bureaucracy*. Philadelphia, PA: Temple University Press.

Garfinkel, H., Lynch, M., & Livingston, E. (1981). The Work of a Discovery Science Construed with Materials from the Optically Discovered Pulsar. *Philosophy of Social Science*, 11, 131-158.

Garson, B. (1988). *The Electronic Sweatshop: How Computers are Transforming the Office of the Future into the Factory of the Past*. New York, NY: Simon and Schuster.

Giddens, A. (1984). *The Constitution of Society: Outline of a Theory of Stratification*. Berkeley, CA: U.C. Press.

Glaser, N., & Strauss, A. (1967). *The Discovery of Grounded Theory: Strategies for Qualitative Research*. Chicago, IL: Aldine.

Goffman, E. (1959). *The Presentation of Self in Everyday Life*. Garden City, NY: Anchor Books.

Goffman, E. (1961). *Asylums: Essays on the Social Situation of Mental Patients and other Inmates*. Garden City, NY: Anchor Books.

Greimas, A.J., & Courtes, J. (1982). *Semiotics and Language: An Analytical Dictionary*. Bloomington, IN: Indiana University Press.

Groneman, C., & Norton, M.B. (Eds.). (1987). *To Toil the Livelong Day: America's Women at Work 1780-1980*. Ithaca, NY: Cornell University Press.

Habermas, J. (1970). *Toward a Rational Society: Student Protest, Science and Politics*. Boston, MA: Beacon Press.

Hartmann, H., Kraut, R., & Tilly, L. (Eds.). (1986). *Computer Chips and Paper Clips*. Washington, D.C.: National Academy Press.

Homans, G. (1950). *The Human Group*. New York, NY: Harcourt, Brace and World.

Hunt, H.A., & Hunt, T.L. (1986). *Clerical Employment and Technological Change*. Kalamazoo, Michigan: W.E. Upjohn Institute for Employment Research.

Jaques, E. (1978). Stratified Depth Theory of Bureaucracy. In Elliot Jaques (Ed.), *Levels of Abstraction in Logic and Human Action: A Theory of Discontinuity in the Structure of Mathematical Logic, Psychological Behavior, and Social Organization* (pp. 209-223). London: Heinemann Education Books Ltd.

Jules-Rosette, B. (1978a). The Politics of Paradigms: Contrasting Theories of Consciousness and Society. *Human Studies*, 1(1), 92-110.

Jules Rossette, B. (1978b). The Veil of Objectivity: Prophecy, Divination and Social Inquiry. *American Anthropologist*, 80(3), 549-570.

Jules-Rosette, B. (1981). *Symbols of Change: Urban Transition in a Zambian Community*. Norwood, NY: Ablex Pub.

Kanter, R. (1977). *Men and Women of the Corporation*. New York, NY: Basic Books.

Klapp, O. (1986). *Overload and Boredom: Essays on the Quality of Life in the Information Society*. New York, NY: Greenwood Press.

Knorr-Cetina, K. (1983). The Ethnographic Study of Scientific Work: Towards a Constructionist Interpretation of Science. In Karin Knorr-Cetina and Michael Mulkay (Eds.), *Science Observed: Perspectives on the Social Study of Science* (pp. 115-140). London: Sage Publications.

Knorr-Cetina, K., & Cicourel, A. (Eds). (1981). *Advances in Social Theory and Methodology: Toward an Integration of Micro-and Macro-sociology.* Boston, MA: Routledge and Kegan Paul.

Latour, B. (1980). Is it Possible to Reconstruct the Research Process?: Sociology of a Brain Peptide. In Karin Knorr-Cetina, Roger Krohn and Richard Whitley (Eds.), *The Social Process of Scientific Investigation: Sociology of Sciences* (Vol. IV), (pp. 53-73). Boston, MA: Routledge and Kegan Paul.

Lynch, M., Livingston, E., & Garfinkel, H. (1983). Temporal Order in Laboratory Work. In Karin Knorr-Cetina and Michael Mulkay (Eds.), *Science Observed: Perspectives on the Social Study of Science* (pp. 205-238). London: Sage Publications.

Marcus, G. (1986). Contemporary Problems of Ethnography in the Modern World System. In James Clifford and George Marcus (Eds.), *Writing Culture: The Poetics and Politics of Ethnography* (pp. 165-193). Berkeley, CA: U.C. Press.

Marcuse, H. (1966). *One Dimensional Man: Studies in the Ideology of Advanced Industrial Society.* Boston, MA: Beacon Press.

Marx, K. (1977/1867) *Capital: A Critique of Political Economy* (Vol. 1) (B. Fowkes, Trans.). New York: Vintage (Original work published in 1867).

Matthaei, J. (1982). *An Economic History of Women in America: Women's Work, the Sexual Division of Labor and the Development of Capitalism.* New York, NY: Schocken Books.

Mayo, E. (1945). *The Social Problems of an Industrial Civilization.* Boston, MA: Harvard University Graduate School of Business Administration.

Mehan, H. (1978). Structuring School Structure. *Harvard Educational Review*, 48(1), 32-64.

Merton, R. (1938). Social Structure and Anomie. *American Sociological Review*, 3(6), 672-682.

Mills, C.W. (1951). *White Collar: The American Middle Classes.* New York, NY: Oxford University Press.

Molstad, C. (1986). Choosing and Coping with Boring Work. *Urban Life*, 15(2), 215-236.

Nakano-Glen, E., & Feldberg, R. (1977). Degraded and Deskilled: The Proletarianization of Clerical Work. *Social Problems,* 25(1), 52-64.

Parsons, T. (1967). An Approach to the Sociology of Knowledge. *Sociological Theory and Modern Society* (pp. 139-165). New York, NY: Free Press.

Perrow, C. (1970). *Organizational Analysis: A Sociological View.* Belmont, CA: Wadsworth.

Peters, T., & Waterman, R., Jr. (1984). *In Search of Excellence: Lesson from America's Best-Run Corporations.* New York, NY: Warner.

Phillips, D. (1974) The Influence of Suggestion on Suicide: Substantive and Theoretical Implications of the Werther Effect. *American Sociological Review*, 39, 340-354.

Roethlisberger, F.J., & Dickson, W. (1966). *Management and the Worker: An Account of a Research Program Conducted by the Western Electric Company. Hawthorne Works, Chicago.* Cambridge, MA: Harvard University Press.

Rotella, E. (1981). *From Home to Office: U.S. Women at Work, 1870-1930.* Ann Arbor, MI: U.M.I. Research Press.

Schachter, S., Ellertson, N., McBride, D., & Gregory, D. (1951). An Experimental Study of Cohesiveness and Productivity. *Human Relations*, 4, 229-238.

References

Schutz, A. (1967). *The Phenomenology of the Social World* (George Walsch and Fiederick Lehnert, Eds., Trans.). Evanston, IL: Northwestern University Press.

Shepard, J. (1971). *Automation and Alienation: A Study of Office and Factory Workers*. Cambridge, MA: M.I.T. Press.

Sheppard, H., & Herrick, N. (1972). *Where Have All the Robots Gone: Worker Dissatisfaction in the 70's*. New York, NY: The Free Press.

Spencer, H. (1886). *Principles of Sociology*. New York, NY: Appleton and Company.

Sperber, D. (1982). *On Anthropological Knowledge: Three Essays*. Cambridge, MA: Cambridge University Press.

Riemer, J. W. (1977). Varieties of Opportunistic Research. *Urban Life*, 5(4), 467-477.

Tyler, S.A. (1985). Ethnography, Intertextuality and the End of Description. *American Journal of Semiotics*, 3(4), 83-98.

Willis, P. (1977). *Learning to Labor: How Working Class Kids Get Working Class Jobs*. New York, NY: Columbia University Press.

Wolf, C. (1986). Legitimation of Oppression: Response and Reflexivity. *Symbolic Interaction*, 9(2), 217-234.

Worthy, J.C. (1950). Organizational Structure and Employee Morale. *American Sociological Review*, 15(2), 169-179.

Yiu Cheong So, A. (1980). What is the Working Class: A Study of the Class Position of Clerical Workers. *Journal of Social Relations*, 8(1), 44-59.